QUANTUM PHYSICS FOR BEGINNERS

The Complete Overview How to Easily Understand the All Principles of Quantum Mechanics in Everyday Life.

James Philips

© **Copyright 2024 by James Philips - All rights reserved.**

The following book is provided below with the aim of delivering information that is as precise and dependable as possible. However, purchasing this book implies an acknowledgment that both the publisher and the author are not experts in the discussed topics, and any recommendations or suggestions contained herein are solely for entertainment purposes. It is advised that professionals be consulted as needed before acting on any endorsed actions.

This statement is considered fair and valid by both the American Bar Association and the Committee of Publishers Association, and it holds legal binding throughout the United States.

Moreover, any transmission, duplication, or reproduction of this work, including specific information, will be deemed an illegal act, regardless of whether it is done electronically or in print. This includes creating secondary or tertiary copies of the work or recorded copies, which are only allowed with the express written consent from the Publisher. All additional rights are reserved.

The information in the following pages is generally considered to be a truthful and accurate account of facts. As such, any negligence, use, or misuse of the information by the reader will result in actions falling solely under their responsibility. There are no scenarios in which the publisher or the original author can be held liable for any difficulties or damages that may occur after undertaking the information described herein.

Additionally, the information in the following pages is intended solely for informational purposes and should be considered as such. As fitting its nature, it is presented without assurance regarding its prolonged validity or interim quality. Mention of trademarks is done without written consent and should not be construed as an endorsement from the trademark holder.

TABLE OF CONTENTS

CHAPTER 1: INTRODUCTION TO QUANTUM MECHANICS   1

1.1 Understanding the Quantum World   1
- Exploring the Quantum Realm   1
- Key Principles of Quantum Mechanics   4
- Historical Development of Quantum Theory   6

1.2 Importance of Quantum Physics in Everyday Life   10
- Quantum Technology and Its Applications   10
- Quantum Computing and Future Innovations   12
- Quantum Entanglement and Communication   15

CHAPTER 2: THE WAVE-PARTICLE DUALITY   19

2.1 Wave-Particle Duality: An Intriguing Phenomenon   19
- Wave Nature of Matter: De Broglie's Hypothesis   19
- Particle Behavior: Photoelectric Effect and Compton Scattering   22
- Observing Duality in Quantum Experiments   24

2.2 Implications of Wave-Particle Duality   28
- Understanding Interference Patterns   28
- Quantum Tunneling and Barrier Penetration   32
- Applications in Electronics and Microscopy   34

CHAPTER 3: QUANTUM SUPERPOSITION AND MEASUREMENT   39

3.1 Concept of Superposition: Schrödinger's Cat Thought Experiment   39
- Understanding Quantum States and Superposition   39
- Experimental Evidence for Superposition   42
- Quantum Interference and Wave Function Collapse   44

3.2 Measurement in Quantum Mechanics   49
- Role of Observers and Measurement Devices   49
- Quantum Zeno Effect and Observational Influence   52
- Practical Implications and Limitations   54

CHAPTER 4: QUANTUM ENTANGLEMENT AND SPOOKY ACTION AT A DISTANCE 58

4.1 Introduction to Quantum Entanglement 58
Definition and Characteristics of Entangled States 58
Bell's Theorem and Violation of Local Realism 60
EPR Paradox: Einstein, Podolsky, and Rosen 65

4.2 Applications of Quantum Entanglement 68
Quantum Cryptography and Secure Communication 68
Quantum Teleportation and Information Transfer 70
Challenges and Future Directions in Entanglement Research 75

CHAPTER 5: QUANTUM MECHANICS IN TECHNOLOGY 78

5.1 Quantum Mechanics in Modern Technology 78
Quantum Sensors and Precision Measurement 78
Quantum Communication Networks 81
Quantum Sensing and Imaging Techniques 83

5.2 Quantum Computing: A Paradigm Shift 87
Principles of Quantum Computing 87
Quantum Algorithms and Computational Advantage 89
Challenges and Opportunities in Quantum Computing 92

CHAPTER 6: QUANTUM PHYSICS AND THE COSMOS 96

6.1 Quantum Cosmology: Exploring the Universe at the Quantum Scale 96
Quantum Origins of the Universe 96
Quantum Gravity and the Search for a Unified Theory 99
Quantum Cosmological Models and Observational Tests 101

6.2 Quantum Effects in Astrophysics and Cosmology 104
Black Holes and Hawking Radiation 104
Quantum Fluctuations and Cosmic Inflation 107
Implications for Understanding the Cosmos 110

CHAPTER 7: PRACTICAL APPLICATIONS OF QUANTUM MECHANICS — 114

7.1 Quantum Mechanics in Everyday Life — 114
Quantum Sensors and Imaging Technologies — 114
Quantum Metrology and Precision Measurement — 117
Quantum Enhanced Materials and Devices — 119

7.2 Future Trends and Emerging Technologies — 123
Quantum Internet and Secure Communication — 123
Quantum Sensors for Medical Diagnostics — 125
Quantum Computing in Finance and Optimization — 127

CHAPTER 8: CHALLENGES AND FUTURE DIRECTIONS — 132

8.1 Current Challenges in Quantum Physics — 132
Overcoming Decoherence and Noise — 132
Scaling Quantum Systems for Practical Applications — 134
Addressing Ethical and Societal Implications — 137

8.2 Future Directions in Quantum Research — 140
Quantum Information Science and Quantum Networks — 140
Advancements in Quantum Computing and Simulation — 142

BONUS CHAPTER: QUANTUM TUNNELLING — 146

find out everything you need to know about quantum tunneling, including how particles can pass through barriers. — 146

SCAN QR CODE TO DOWNLOAD FREE AUDIOBOOK VERSION — 150

CHAPTER 1: INTRODUCTION TO QUANTUM MECHANICS

1.1 UNDERSTANDING THE QUANTUM WORLD

EXPLORING THE QUANTUM REALM

As we scoot through our digital era at what feels like the speed of light, there's an invisible giant shaping our everyday lives in ways that might seem the stuff of science fiction, and that giant is quantum technology. You've probably heard the term strewn about in tech circles or popping up in the latest smartphones feature list. But what does it truly entail, and how does it reach beyond laboratory walls into the palm of your hand, the glint of your watch, and the future cities you dream of?

Quantum technology, my friends, rests upon the quirky behaviors of nature's smallest constituents—particles and their quantum states. These states, which encompass properties like position, momentum, and spin, behave unlike anything we experience in our 'classical' world. In the realm of the quantum, particles can exist in multiple states at once; they can be intertwined across vast distances, and they often defy classical logic and notions of causality.

Now, let's break the ice with quantum's most entrancing application: quantum computing. Here's something to chew on: classical computers, which power everything from your microwave oven to the International Space Station, calculate using bits. These bits are like tiny on-off switches that can either be in position **0** or **1**. But quantum computers use qubits, which can be both **0** and **1** simultaneously, thanks to superposition. Imagine being able to read and turn pages of multiple books at once. Qubits extend this incredible ability to process vast arrays of data simultaneously, meaning quantum computers can tackle problems in minutes that would take traditional computers millennia.

Consider encryption and cybersecurity. Presently, our data is protected by complex cryptographic systems, which, given enough time or computational power, can theoretically be broken. Quantum cryptography, however, harnesses the laws of physics—like the impossibility of measuring quantum states without altering them—to create theoretically unbreakable encryption. Not just secure, but underpinned by the fundamental laws of the universe. Talk about a step-up in privacy!

Next up is medicine, where quantum technology promises to revolutionize how we detect and understand diseases. Imaging techniques magnified by quantum entanglement are set to offer unprecedented views of the intricacies within our cells, perhaps even leading us to early cancer detectors that work with the gentlest whisper of a signal. Moreover, quantum computers' capacity to simulate molecules' behavior heralds a new dawn for drug discovery, potentially accelerating the development of life-saving medicine manifold.

But if you're picturing quantum tech as some distant utopia, hold on to your hats. Today, precise quantum sensors are roaming among us, improving navigation systems beyond the reach of even the most accurate GPS—and they're not scared to venture off the beaten path. We're talking about navigation solutions for places where GPS can't penetrate—submarines delve into ocean depths, and spacecraft venture into cosmic expanses—all guided by the delicate dance of quantum particles.

Everywhere you look, from the environmental field to finance, quantum technology is becoming an ace up the sleeve. Climate scientists now have their hands on quantum-enhanced models that trawl through colossal sets of variables to predict weather patterns with jaw-dropping accuracy. In finance, powerful quantum algorithms chew through market data at a staggering pace allowing for optimization of investments and risk analysis at speeds and depths previously unimaginable.

And let's not forget about energy; in the quest to create more efficient solar panels, researchers are taking a quantum leap. They're delving into the quantum properties of molecules to usher in materials that could capture more sunlight and convert it into electricity, offering a beacon of hope for clean energy in an increasingly thirsty world.

Let's be honest, though—embarking on the quantum journey isn't without its snags. For all its potential, quantum tech still clambers through infancy. Quantum systems are notoriously delicate, and maintaining the coherence necessary for these particles to perform their magical feats requires a dance with absolute precision. Moreover, as these technologies advance, ethical considerations balloon. How do we wield such powerful tools responsibly? How do we ensure quantum's benefits aren't hoarded but shared across the globe?

Yet, despite these challenges, the prospect that lies ahead is nothing short of a scientific renaissance. An era where diseases are not just cured but prevented, where communication is secure in a way no cyber-attack can compromise, where our understanding of the universe expands exponentially through quantum lenses.

As we orchestrate the symphony of this quantum awakening, what remains most striking is the notion that these infinitesimal particles, with their paradoxical laws, don't just inhabit some esoteric realm of physicists. They're here, they're real, and they're unfolding the tapestry of a future that intertwines the fantastic with the ordinary. A future that you, dear reader, are already becoming a part of.

Sometimes the quantum world might seem like a dreamscape, divorced from the routine rhythm of daily life. But the reality couldn't be more different. Every time you send a secure message, check the weather forecast, or dream of a future untouched by the scathing breath of fossil fuels, you're witnessing the shimmering hints of quantum physics' powerful aplomb.

Through the language of quantum mechanics, we step into a universe where the small is mighty, and the invisible is transformative. This is the quantum world's calling card—an open invitation to envision a life interwoven with wonders born out of the heart of the atom, wonders that are redrawn our present and are poised to redefine our future.

KEY PRINCIPLES OF QUANTUM MECHANICS

Delving into the enthralling sphere of quantum physics often brings to mind esoteric phenomena that seem light-years removed from the humdrum of our daily lives. Yet, one of the most exhilarating inroads—a field boasting both mind-bending principles and pragmatic allure—is quantum computing. Far more than a mere upgrade to the computers that we rely on today, quantum computing propels us into a future rich with innovations that we're beginning to fathom only now.

Imagine, if you will, a world where diseases are conquered faster because we can simulate molecular structures in unprecedented detail, or where financial markets are analyzed with such precision that economic crises become relics of the past. This world inches closer to reality daily, fueled by the promise of quantum computing. The essence of its power lies not in faster or larger-scale computations of the kind we're accustomed to, but rather in its ability to process a multitude of possibilities simultaneously.

The Quantum Leap in Computing

Quantum computing diverges from classical computing at the fundamental level of information representation. Classical computers encode information as bits—a binary sequence of **0**s and **1**s. In stark contrast, quantum computers employ quantum bits or 'qubits.' A qubit can exist not only in a state corresponding to the logical state **0** or **1** but also in states involving the superposition of these classical states. Think of this as the qubit's ability to embody multiple states at once, until it is measured.

This attribute alone propels quantum computers into a realm of staggering computational parallelism. For every additional qubit that a quantum computer possesses, its processing power doubles, unlocking capabilities beyond the reach of classical computing. The difference is so profound that problems deemed intractable for even the mightiest supercomputers may soon yield to the might of quantum processors.

Bridging Quantum Computing and Everyday Innovation

The repercussions of this quantum leap will reverberate across all sectors, seeding future innovations. Indeed, quantum computing carries the promise to totally transmute industries by cracking complex optimization problems—be it in logistics to streamline supply chains or in AI algorithms that adapt in real-time to evolving data.

In the realm of cryptography, quantum computers present a double-edged sword. On one hand, they could shatter the security of current encryption schemes, necessitating a surge towards quantum-resistant cryptography. On the other hand, they herald the dawn of what is touted as 'quantum cryptography'—a secure communication system so robust that it is protected by the very laws of physics.

One should envision a future where weather forecasts are so precise, they allow cities to prepare for disasters with accuracy never before possible, thanks to quantum computers' ability to model complex systems. Drug discovery and materials science stand on the cusp of a revolution, as quantum simulation offers insights into molecular and quantum systems that were merely hypothetical before.

Quantum Computing's Challenges: An Honest Look

While the visionary projections of quantum computing's capabilities can captivate any technophile, the path to realizing these marvels is laden with obstacles. For one, maintaining the integrity of qubits is a feat in itself. Qubits are delicate, susceptible to the slightest environmental perturbations—this phenomenon is referred to as 'decoherence.' Preserving their 'quantumness' long enough to perform complex calculations is a dance on the edge of technological possibility.

There's also the challenge of qubit interconnectivity and scalability. To perform the elaborate dance of algorithms, qubits must entangle with precision across vast quantum networks. Building a quantum computer with a significant number of qubits that can maintain entanglement is our generation's space race—a journey of exhilarating potential fraught with scientific and engineering hurdles.

Quantum Computing: A Tantalizing Horizon

Yet, for all these challenges, the future of quantum computing is tantalizingly within reach. The rapid pace of research means that what was once theoretical is increasingly turning practical. Companies and governments are pouring immense resources into quantum computing, heralding a high-stakes technological competition akin to the digital age's inception.

In the realms of academia and industry, collaborations are flourishing to pinpoint the most effective algorithms and protocols through which quantum advantage—that magical tipping point where quantum computing unequivocally outstrips classical computing—can be realized. The race is on, not just in laying down qubits but in crafting the very fabric of the future's quantum computing landscape.

Embracing a Quantum Future

As we stand on this pivotal curvature in the story of our technological evolution, it's clear that quantum computing is not a distant shimmer but an approaching dawn. As curious minds seeking to understand our world, quantum physics proves not just theoretically fascinating but indispensable in underpinning future technologies that will reshape our very notion of what is achievable.

You, the reader, whether from a background in computer science, engineering, or simply mesmerized by the bountiful prospects technology opens before us, are witness to this transformative epoch. Quantum computing is more than an esoteric branch of physics; it is the foundation upon which we will build previously unthinkable innovations. The quantum age beckons, offering a fresh canvas for the intellectually bold to reimagine and reinvent our everyday existence. As we make the quantum leap, it's not just quantum bits at stake; it's our global future, bursting with potential, awaiting the intrepid and the insightful.

HISTORICAL DEVELOPMENT OF QUANTUM THEORY

Imagine for a moment you're in a silent auditorium, waiting for a performance to begin. The stage is set for one of the most mystifying acts ever to be encountered in both science and the daily dance of particles coursing through the veins of modern technology, an act known as quantum entanglement. In the grand theater of quantum physics, it is both a phenomenon and a backbone for revolutionary communication methods that are quickly stepping out of science fiction into the world we live in.

Quantum entanglement is a term that might evoke images of a magician's intricate cords knotting and unknotting with inexplicable precision. Much like these cords, entangled particles remain connected so that the state of one instantaneously influences the state of another, no matter how vast the distance between them. This peculiarity is not just a quirk of the microscopic world; it harbors profound implications for how we communicate across the globe and beyond.

Entanglement begins on the stage of the small, amid particles that we are unable to see with the naked eye. When two particles become entwined in the quantum tango of entanglement, they become so deeply linked that the properties of one directly correlate with the properties of the other. Alter one particle, and its distant partner mirrors this change immediately. It's a bond that Einstein famously referred to as "spooky action at a distance," yet this spine-tingling event is not just a theoretical fancy. It is experimentally proven, repeatedly displayed in laboratories worldwide.

One may wonder—does this mean we can achieve instantaneous communication? Should I cast away my smartphone in anticipation of a quantum device capable of conveying messages faster than the speed of light? There's a caveat. While entanglement does allow for particles to be linked across distances, it is subject to the no-communication theorem. This principle dictates that, though entangled particles are correlated in their states, they can't actually send messages in the conventional sense. The moment we attempt to observe or use a particle to convey a message, the entanglement collapses. Thus, the dream of faster-than-light chit-chat remains just that—a dream.

However, quantum entanglement is not stripped of its potential in the communication landscape. The magic it does offer comes in the form of quantum cryptography. This is where quantum entanglement shows its true, practical might. Quantum cryptography utilizes entangled particles to create an unbreakable code. Any attempt at eavesdropping disturbs the entangled particles and can be immediately detected by the lawful communicators. This novel approach promises a future where privacy is not a mere suggestion but an enforceable law of the universe.

From financial transactions to confidential communications, quantum cryptography is set to redefine what we understand by the word 'secure'. It's akin to having an envelope that self-shreds the moment someone unwarranted lays a finger on its seal. This is not just wishful thinking; it's a technology under active development. Companies and governments are pouring resources into making quantum communication networks a reality—a testament to the importance and transformative power of quantum mechanics in our interconnected world.

Yet, there is more. Quantum communication is gradually stepping out of the realm of security and into broader territories. For instance, quantum networks enable physicists to create 'quantum repeaters' which facilitate long-range quantum communication without the need for a continuous medium. It is much like passing a message along in a game of 'telegraphic whispers,' where despite gargantuan distances, the original phrase reaches the end without a single alteration.

Pushing the boundaries further, satellite-based quantum communication employs the bizarre features of quantum mechanics to link earthbound stations with devices in orbit. The Chinese satellite Micius, for example, has successfully demonstrated entanglement over thousands of kilometers. This endeavor is not just a proof of concept; it is a seed from which a global, unhackable communication system could sprout.

As history has often shown, with new technology comes great responsibility. Quantum entanglement and the communication paradigms it enables require us to rethink our approaches to information security, privacy, and perhaps even the structure of the internet as we know it. We are at the cusp of a quantum revolution where the melding of technology and the most profound aspects of nature promises to catapult our communication capabilities into hitherto uncharted territories.

So, does all of this imply that you, sitting with your morning coffee and scanning the news on your tablet, are on the brink of experiencing quantum communication in your everyday life? Possibly not today or tomorrow, but the wheels are in motion. The trickle-down effect from high-end, secure government transmissions to the everyday text message is historically inevitable.

Much like the ripple effects from the invention of the semiconductor, the quantum leap into entanglement-based communication will influence domains far from the microscopic origin of this extraordinary phenomenon. Consider health care, where secure transfer of medical records could mean the difference between life and death, or governance, where the sanctity of ballots rests on the integrity of quantum mechanisms.

The curtain rises on a world in which quantum entanglement and communication stand poised to revolutionize the bedrock of human connections. It's a magisterial world that faithfully honors the intricate choreography of the tiniest actors in the quantum realm for the grand drama of our day-to-day lives. Your role in this unfolding act? Remain curious, informed, and ready for a future where once-spooky action at a distance becomes another dependable thread in the fabric of daily communication.

1.2 IMPORTANCE OF QUANTUM PHYSICS IN EVERYDAY LIFE

QUANTUM TECHNOLOGY AND ITS APPLICATIONS

Step into the quantum universe—it's a realm so radically different from our everyday experiences that it can feel like an alternate reality. In this wonderland, particles can be in several places at once, invisible connections span vast distances, and the mere act of watching can alter the outcome of events. But fret not; even the most bizarre phenomena can be made accessible with the right guide. Let's delve into the captivating world of quantum entanglement and discover how this eerie concept is not just a philosopher's fancy but a cornerstone of modern technology.

At its heart, quantum entanglement is a peculiar connection between particles. These ties are so powerful that the state of one entangled particle instantly influences its partner, irrespective of the distance separating them. This strange synchrony defies the classical view of the world, where things influence each other through direct interactions or at a pace limited by the speed of light. Instead, entangled particles act as twin wizards with an uncanny, unspoken understanding—alter one, and you instantaneously tweak the other.

Pioneered by Einstein, Podolsky, and Rosen in **1935**, entanglement was a way to challenge quantum mechanics. Einstein famously dubbed it "spooky action at a distance," underscoring his discomfort with the idea that particles could exert influence without a connecting force detectable by the physics of the day. This revelation quickly became not just a curiosity, but a pivotal moment in physics—posing deep questions about the nature of reality and locality.

Testing the limits of entanglement falls to Bell's Theorem. In the **1960s**, physicist John Bell formulated an empirical test to determine if hidden, unaccounted factors could be responsible for the spooky coordination seen in quantum entanglement. The results? Quantum mechanics passed with flying colors, leaving the scientific community with irrefutable proof that the entangled states are real, albeit profoundly enigmatic.

Let's consider the applications, shall we? Quantum entanglement has left the confines of thought experiments and is now being harnessed in cutting-edge technologies. One of the most promising applications is in quantum cryptography—the art of sending secure messages. In this futuristic encryption, a key encoded in the quantum states of particles is used to lock away information. If an interloper tries to eavesdrop, the entangled particles would react and betray the spy's presence, thus keeping the communication secure.

Another innovation inspired by entanglement is quantum teleportation. While it may not yet be the stuff of science fiction, teleporting information at the quantum level is very much a reality. Here, we exploit the connection between entangled particles to transmit the state of one particle to another, over potentially vast distances, without sending the particle itself. This breakthrough has vast implications for quantum computing and constructing a quantum internet—a network that leverages quantum characteristics to enable incredibly powerful and secure data sharing.

Though the boon of quantum entanglement is undeniable, the challenges looming on the horizon are substantial. Building systems that maintain entanglement over long distances and durations—a necessity for practical applications is no small feat. Entangled particles are sensitive sorts; they don't take kindly to interference and can lose their connection, a phenomenon known as decoherence, in the presence of noise and external factors. Current research is feverishly addressing this, looking to develop materials and methods to protect these fragile quantum states from the ravages of their environment.

And now, to the mind-bending implications of entanglement for the fabric of reality. This element of quantum physics suggests a level of interconnectivity in the cosmos that is not just bizarre, but beautiful. Perhaps the most far-reaching implication of entanglement is the idea that it calls into question the very notion of separateness—illustrating that at a fundamental level, the universe might be an intricately interwoven tapestry, with each thread invisibly connected to countless others.

To gaze upon the entangled particles is to gaze upon the cosmos from a fresh perspective. Physicists and philosophers alike are challenged to rethink the traditional notions of space and causality. The entanglement posits a world where distances are trivial at the quantum level, where cause and effect might intertwine in ways not yet fully comprehended, and where our understanding of time itself could be revolutionized.

Even as we stand marveling at the foot of Mount Quantum, it is essential to acknowledge that entanglement, with all its potential, is still shrouded in mystery. Researchers continue to probe the depths of this phenomenon, not only to harness it for technological progress but also to uncover the deeper truths it holds about the universe.

As we close this glimpse into quantum entanglement, remember that these quantum quirks are not mere abstractions—rather, they are part of the very foundation that shapes our existence and the universe at large. With every advancement, the promise of quantum entanglement nudges a little closer, not just to the realms of possibility, but inevitability. Yes, the path ahead has its share of hurdles, yet it also brims with opportunities to witness—and perhaps take part in—a profound transformation in our understanding of reality itself. Embrace the journey, for each step reveals wonders that, like the entangled particles, are bound to leave you forever changed.

QUANTUM COMPUTING AND FUTURE INNOVATIONS

When you gaze up into the star-speckled night, you're not just looking at celestial bodies hundreds, thousands, or even millions of light-years away. You're peeking into a cosmic dance choreographed by the principles of quantum mechanics, painted on the grand canvas of the universe. At this moment, quantum mechanics isn't just an abstract set of ideas; it's the artist behind the brushstrokes of reality as we know it.

Let's shrink down to the quantum scale, where the cosmos we are familiar with is born, breathes, and evolves. Quantum cosmology is the branch of physics that uses quantum mechanics to explain the very early universe, particularly during the Planck epoch, moments after the infamous Big Bang. During this time, the universe's size was comparable to that of a quantum particle, and accordingly, it was subject to quantum laws.

Our standard model of cosmology tells us that about **13.8** billion years ago, the universe was condensed into a singularity, an infinitesimal point where the known laws of physics cease to function. This Big Bang singularity springboarded our universe into existence. However, here's a little quantum twist: when we attempt to peer back in time, close to the origin point, our classical physics narratives crumble. We need something more—something that gracefully dances with both the quantum and gravitational realms. Tuned into Planck-scale phenomena, quantum cosmology aims to sing a harmonious prelude to the Big Bang, fixing the cacophony wrought by general relativity when pushed to its limits.

The seeds of cosmic structure we see today — galaxies, stars, and planets — originated from quantum fluctuations, minuscule variations in the density of the universe's mass-energy. Traditionally, these irregularities would seem insignificant, but through the lens of quantum mechanics, we understand that they're colossal. The inflationary model of cosmology suggests that, due to cosmic inflation, these fluctuations were stretched to macroscopic scales, eventually serving as the scaffolds for large-scale structures like galaxy clusters.

Then there's the elusive enigma of dark matter and dark energy. These mysterious components dictate the universe's expansion and structure and yet seem untethered to our conventional understanding of matter. The quantum universe is home to an array of exotic particles, some of which might just hold the keys to this cosmic mystery. As astrophysicists and quantum physicists tango, the boundaries blur, and the prospects of unveiling the true nature of dark matter and energy draw closer with each quantum revelation.

Black holes, those gravitational monstrosities where not even light can escape, grip our imaginations. They are fascinating not just because of their sheer power and mystery but also due to their quantum conundrum, Hawking radiation. Pioneered by Stephen Hawking, the theory postulates that black holes aren't the eternal prisons we once thought. Quantum effects near the event horizon enable particles and antiparticles to pop into existence, with some escaping as Hawking radiation. This startling insight suggests that black holes could eventually evaporate, dripping into the cosmos through quantum leaks.

Now, pondering the interplay of quantum mechanics and gravity naturally leads us to the quest for a unified theory, a 'Theory of Everything.' The Holy Grail of physics, if you will. Quantum gravity, this hypothetical framework, would marry general relativity with quantum mechanics, offering a consistent description of all four fundamental interactions in the universe. Numerous paths are being forged in this quest, string theory and loop quantum gravity being the frontrunners. While these ideas are beautifully intricate, resembling a symphony of the smallest violin strings and the weavings of spacetime loops, they are not without their critics. Producing testable predictions to confirm these theories remains one of the most formidable challenges science faces today.

In the intricate ballet of quantum cosmology, we can't omit cosmic inflation theory. Pardon the pun, but it expanded our understanding of the universe's birth—literally. Inflation posits that in the first split seconds of existence, the universe underwent an explosive expansion faster than the speed of light, driven by a scalar field known as the inflaton. This fleeting period ironed out any wrinkles in spacetime, leading to the smooth and homogeneous universe we observe on large scales. While inflation is widely accepted, it's not without its skeptics and foes. The theory is still the seedbed for heated discussions and tantalizing theories among cosmologists and quantum theorists.

Quantum mechanics also ushers us into otherworldly landscapes, where the concepts of space and time wash away as waves and particles intermingle. The smallest fluctuations here do not merely vanish; they become the architects for the creation of potential universes, a multiverse if you will. This is where quantum cosmology propels our imagination beyond the edge of science fiction, into mind-bending realms where our universe could be just one of an unfathomable number of bubbles in a cosmic foam.

The narrative of quantum cosmology is not merely a collection of theoretical musings. Using advanced equipment like the Laser Interferometer Gravitational-Wave Observatory (LIGO) and the Planck space observatory, scientists are rigorously testing these theories. They're looking for imprints of gravitational waves, hints of inflation, signatures of quantum fluctuations, and shadows of dark matter. Each finding audaciously ticks like a cosmic clock, resonating with the rhythm of quantum beats.

The journey of quantum cosmology is, without a doubt, an ambitious voyage through the most profound questions of existence. We are barely scratching the surface, and while the map is still sketchy, the compass of quantum mechanics keeps us astern in the turbulent seas of the cosmic unknown. Facing this frontier, we should tread with insatiable curiosity, humility, and awe at the splendor unfolding before us. After all, quantum mechanics and cosmology together draw the outlines of our cosmic storybook—one where we're both authors and readers, eager to turn the next page.

QUANTUM ENTANGLEMENT AND COMMUNICATION

Imagine you're at the seashore, watching waves tumble and crash, a rhythmic dance of energy and motion. Now, transport that image to the far smaller, surreal landscape inside the atom, and you'll have stumbled upon one of the most brilliant and baffling ideas in physics: de Broglie's hypothesis of wave-particle duality. This is the window into the quantum world, where the rules of classical mechanics lose their grip and a universe of possibilities opens up.

In the heartland of the twenties, a time when quantum mechanics was just beginning to shape our understanding of the nanoscopic realms, Louis de Broglie, a French physicist, made a proposal that'd challenge the very nature of matter. Borrowing from the dualistic nature of light, which had already been shown to exhibit both wave-like and particle-like behaviors, de Broglie suggested that these dual properties were not the exclusive property of photons. Every particle, he posited, could also behave like a wave.

This revelation came with a simple, yet profound equation:

$$\lambda = \frac{h}{mv}$$

Here, (λ) represents the wavelength (the wave aspect of matter), (h) is Planck's constant, (m) stands for mass, and (v) the velocity of the particle. With this, de Broglie opened our eyes to a universe where every particle has its own wave, a 'matter wave', and the scale of these waves varies inversely with momentum.

Let's unpack what this means. Imagine a baseball hurtling through the air. The ball's mass and speed together give it a substantial momentum; according to de Broglie's equation, its wavelength would be incredibly tiny, far too small to observe any wave behavior. For all intents and purposes, the baseball remains purely particle-like in our everyday experience.

Now shrink down to the subatomic particles, like electrons zipping around an atom. Their mass is minuscule, and even though they're moving at high speeds, their momentum is still relatively small. This results in a larger wavelength, one which is comparable to their size. Here, in the quantum realm, the wave aspect of particles becomes a match for their particle nature, and we can no longer neglect it.

In **1927**, the validity of de Broglie's hypothesis received a resounding affirmation from Clinton Davisson and Lester Germer. They conducted experiments by firing electrons at a nickel crystal and discovered that the electrons scattered in a manner that was characteristic of wave interference, creating a diffraction pattern akin to what you'd expect from waves interacting with a barrier containing slits, as is the case with light.

The ramification of this is both thrilling and puzzling. It implies that before an electron hits a detector, it travels not in a straight line, but as a spreading wave, capable of taking many paths simultaneously. When the wave reaches the detector, the 'probability wave' collapses into a particle, but before that, the electron exists in a blurry, uncertain state, spread out over space.

But the universe doesn't let us in on the trick; we can only detect the particle-like part of the electron when it's measured. Before measurement, we see the world through the eyes of waves, and all particles engage in a cosmic dance of possibility. Where an electron is when it's not being watched is a question without a definitive answer; it exists in a form that's not entirely a wave nor entirely a particle.

All of this can feel a little vertigo-inducing. If you're starting to feel that the ground under your feet isn't as solid as you thought, you're not alone. But de Broglie's hypothesis is not just an esoteric idea with philosophical implications; it has practical applications that have revolutionized science and technology.

It's what makes everything electronic around us work. Think of semiconductors – the backbone of all modern electronics, from your smartphone to the internet. They are designed with an intimate understanding of the wave nature of matter. Thanks to de Broglie, we can engineer materials at the atomic level to have specific electrical properties. These precise manipulations are what give us the control needed to build transistors, the tiny switches that form the bedrock of all digital circuits.

Furthermore, the concept of wave-particle duality leads to the powerful technique of electron microscopy, allowing us to achieve images with a resolution order of magnitude greater than with conventional light microscopy. By using electrons and acknowledging their wave nature, we can peer into the structure of materials down to the atomic scale, unlocking a universe previously concealed from the naked eye.

What we've covered is just the tip of the quantum iceberg. The leap from de Broglie's hypothesis to today's applications is enormous and requires a bridge of many scientific milestones. Yet, the starting point of that journey is this beautiful, elegant but starkly challenging principle, which tells us that at the fundamental level, the distinctions between waves and particles blur and the nature of reality becomes a mysterious, enveloping wave of probabilities.

As we continue to explore the quantum realm, expect to witness waves of matter shaping the very fabric of technology and perhaps, the future itself. The wave nature of matter, once just a hypothesis, now stands as a cornerstone of our understanding in quantum mechanics and the vast ocean of innovation it holds. It's a testament to the power of human curiosity and ingenuity that we've come this far, unraveling the quantum symphony scored by nature itself. And with each day, each new discovery, the symphony grows richer, the dance of particles and waves more complex, choreographed by the fundamental laws that we're only beginning to comprehend.

CHAPTER 2: THE WAVE-PARTICLE DUALITY

2.1 WAVE-PARTICLE DUALITY: AN INTRIGUING PHENOMENON

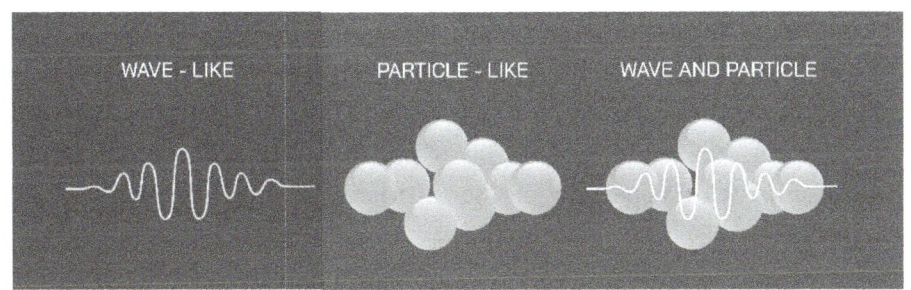

WAVE NATURE OF MATTER: DE BROGLIE'S HYPOTHESIS

Imagine taking a leisurely stroll through a tranquil garden, the sort bathed in sunlight, with a gentle stream trickling through. You toss a pebble into the water, and ripples spread outwards, intersecting and creating a pattern of troughs and crests that dance upon the surface. Such is the picturesque theatre for one of quantum mechanics' most enthralling acts: interference patterns. This phenomenon gifts us with a profound insight into the wave-particle duality, revealing the quixotic nature of matter at the quantum level.

Interference patterns emerge when waves collide and merge, weaving a tapestry of alternating light and dark bands or rings. In our quantum realm, these patterns provide a dazzling visual confirmation of the wave-like properties of particles, akin to the ripples in our garden stream. But rather than being solely a spectacle of beauty, interference has vast implications throughout quantum mechanics and beyond, inviting us to explore its subtleties and applications.

The most iconic demonstration of this behavior is through the famous Double-Slit Experiment, first performed with light, and later, audaciously, with electrons. This experiment showcases the mesmerizing wave-like character of particles. When emitted towards a barrier with two slits, each particle passes through not as a single entity, but spread out like a wave, traversing both slits simultaneously and interfering with itself on the other side. The particles, wave-like phantoms, congregate on a detection screen in lines of peaks and troughs, creating a striated pattern of presence and absence – the interference pattern.

But our story grows even more curious. If you attempt to unveil the particle's path by observing which slit it travels through, the pattern transforms. The act of measurement collapses the wave function – a peculiar term we use to describe the quantum state of a particle. The pattern on the screen shifts from an interference pattern to one resembling bullet shot through a single slit, as though the particles were shy performers that change their act when watched. This illustrates the paradoxical duality of particles, acting as both waves and point-like particles, depending on the experimental setup and whether they are observed.

This cosmic ballet of quantum particles holds the keys to unlocking the behavior of the material world at its most fundamental. The interference patterns are not merely for photogenic appeal; they are the brushstrokes that paint our understanding of the quantum world.

But let us now pivot from these ethereal descriptions to concrete realities. What practical applications do these airy quantum interference patterns hold within their folds?

In the world of electronics, these principles are harnessed in devices like transistors, integral to the operation of everything electronic that we hold dear. Quantum tunneling – where particles pass through barriers they seemingly shouldn't – is directly related to interference effects. The delicate interplay between wavefunctions determines the likelihood of a particle "tunneling" through an obstacle, giving life to technologies that would dazzle the scientists who first charted these quantum shores.

In microscopy, the precision granted by quantum interference allows us to visualize the atomic fabric of materials with electron microscopes. By exploiting wave interference patterns, these microscopes achieve unparalleled resolution, revealing the intricate architectures of molecules and materials – an indispensable tool in fields as diverse as material science, biology, and nanotechnology.

For those passionate about the heavens, quantum interference patterns help astronomers and physicists peer into the universe's enigmatic heart. Devices known as gravitational wave detectors employ the concept to reveal ripples in spacetime – gravitational waves, those elusive messengers from cataclysmic cosmic events such as black hole collisions.

And for those seeking to keep their virtual lives secure, interference patterns provide the bedrock for quantum cryptography systems. The delicate nature of quantum states, susceptible to the faintest disturbances, enables the creation of virtually unbreakable cryptographic codes. An eavesdropper's slightest attempt to intercept this quantum communication disturbs the interference pattern, revealing the intrusion.

We are on the cusp of a revolution, with quantum interference patterns as both muse and architect. The burgeoning field of quantum computing relies on the twin pillars of superposition and interference to solve problems that traditional computers would take millennia to crack.

Interference patterns also provide a canvas for exploring the most fundamental inquiries of quantum mechanics. They are the frontline in probing the delicate coexistence of quantum coherence and decoherence – the ability of a quantum system to maintain superpositions versus their propensity to lose this ability and behave classically under certain conditions.

The hope of discovering new physics lies within these patterns, as subtle departures from expected results could signal unknown forces or particles, beckoning us towards an even deeper understanding of the universe.

As we stand back and marvel at the implications of wave-particle duality and the interference patterns it creates, we find ourselves utterly enchanted. The quantum world, so often cloaked in abstract mathematical language, extends a hand through these patterns – inviting us to experience its wonders and explore its untold potentialities.

So let us continue our journey through the quantum landscape with a curious mind and open heart, taking with us the lesson of the interference patterns: that in complexity and elegance lies an intricate story—a story of nature's deepest truths, sculpted by the principles of quantum mechanics. Whether it be in the intricate design of a technological marvel, the encrypted messages zipping through a future quantum internet, or the hum of the cosmos itself, these patterns play a starring role in the quantum symphony that underpins our reality. As we turn the page, we eagerly anticipate the next act in this mesmerizing exploration of the quantum world.

Particle Behavior: Photoelectric Effect and Compton Scattering

In the mesmerizing arena of quantum mechanics, where particles waltz to the tune of probabilities and defy our classical expectations, there exists a phenomenon as beguiling as a magic trick - quantum tunneling. Imagine you are a tiny subatomic particle, say an electron, facing an energy barrier—a wall too high for you to leap over by classical standards. What do you do? Surrender to the impenetrable obstacle? Not in the quantum realm. Instead, you might do the unthinkable: you materialize on the other side!

That's quantum tunneling for you, a peculiar consequence of the wave-particle duality that encapsulates the quantum world's maverick nature. In classical physics, a ball doesn't roll over a hill if it doesn't have enough energy to surmount it. However, quantum objects are part probability wave, and these waves can extend beyond the confines of the barrier, granting the object a non-zero chance to appear on the other side. This sounds like something straight out of a science fiction story, doesn't it? Well, this is the awe-inspiring truth of our universe at its smallest scales.

For instance, take the scanning tunneling microscope (STM). It leverages quantum tunneling to create images of individual atoms on a material's surface—the peaks and valleys of the atomic landscape rendered visible by electrons that defy barriers. Then there's the flash of life intermediary in the process of nuclear fusion within stars. Protons, bathed in the fiery cauldron of stars, are able to merge due to tunneling—quantum mechanics at play in the celestial forge. And that's just skimming the surface; tunneling electrifies the field of electronics too. Modern transistors, the heart of your computer's processors, rely on controlled tunneling to switch on and off billions of times per second.

Quantum tunneling isn't just a curious quirk; it unveils a universe that's vibrantly non-intuitive, whispering promises of technologies yet to be envisioned. The phenomenon arises from the marriage of two foundational principles: the Heisenberg Uncertainty Principle and the probabilistic nature of quantum mechanics. The Uncertainty Principle stipulates that we cannot know both the position and momentum of a particle with absolute certainty. Thus, a particle approaching a barrier has an intrinsic fuzziness in its location, a haze of probability where it can be here, there, or—the moment of 'aha!'—beyond the barrier.

As you delve into the wave function describing the quantum state of our particle, you'll see it doesn't just abruptly end at the barrier but instead, it trickles through, albeit with rapidly decreasing amplitude. This penetrating tail embodies the ghost of a chance that the particle will defy the odds and pop up on the other side—a quantum sneeze that can't contain itself.

Now, let's stretch our minds further to grasp the influence of the barrier's characteristics on this tunneling. The thickness and energy height of the barrier affect the tunneling probability, a relationship governed by a wave's behavior when encountering obstacles. A thinner or lower barrier makes the probability of tunneling higher; meanwhile, a gargantuan, energy-dense wall would render tunneling an event so rare it might as well waltz into the realm of legend.

Strikingly, tunneling has gifted us with the development of the tunnel diode, a semiconductor device which operates on high-speed quantum tunneling to switch currents. Its swift reaction time has clutched its place in high-frequency oscillators and fast digital circuits. Furthermore, the notion of quantum tunneling is seeping into the exploration of more advanced quantum dot technology, stitching the future of electronics with the unwavering thread of quantum phenomena.

But let's plant our feet back on Earth for a moment—can we use quantum tunneling in our everyday technology? Most certainly. Consider flash memory, which stores the cherished memories of your digital photos. It exploits quantum tunneling to trap or release electrons, encoding the bits that hinge on our collective digital legacy. Or ponder the vast potential spawning from tiny organic molecules that harvest the sun's bounty; their competence in energy conversion is, in part, due to electrons quantum tunneling across molecular junctions.

As we revel in the potential of quantum tunneling, it's prudent to acknowledge that it also holds sway over one of the most grandiose mysteries—the fate of our universe. "Vacuum decay" is a theoretical scenario where quantum tunneling could prompt the universe to transition to a lower energy state, toppling the cosmos into a new form. It is, thankfully, a hypothetical musing rather than a clear and present danger.

Make no mistake, the journey to comprehensively harnessing quantum tunneling is strewn with challenges. Take decoherence, the nemesis of quantum systems, where the purity of a quantum state is spoiled by the chaos of the environment. Efforts to precisely navigate tunneling events must wrestle with this disruptive influence to unlock the full potential of quantum technologies.

Let's pause to marvel at this quantum odyssey. The principles once cloaked in impenetrable mystery are steadily unfolding to unlock doors we never knew existed. Quantum tunneling isn't just a fascinating scientific spectacle; it's a voyager charting the path to a future rippled with quantum possibilities. From enhancing the speed of our computing machinery to poking at the threads of reality itself, it's a clear reflection of the power and subtlety wrapped up in the quantum wonderland—a realm where barriers are simply invitations to innovate.

Observing Duality in Quantum Experiments

Wave-particle duality, the bedrock of quantum mechanics, presents us with a startling truth: every quantum entity, from photons to electrons, embodies both wave-like and particle-like properties. This Schrödinger's Box of the quantum world, where entities exist in a state of duality, has ramifications that extend far beyond theoretical wonderment. It's a key that unlocks a myriad of applications in the realms of electronics and microscopy, technologies that profoundly animate modern life. Our voyage into these practical manifestations uncovers an elegant fusion of quantum abstraction with tangible innovation.

The heartbeats of our digital age lie in the silicon pathways of electronic devices. Picture the traditional transistor, acting as a binary switch within the processors of our computers, an emblem of the physical principles governing classical electronics. Yet, the quantum leap occurs when wave-particle duality takes center stage. Electrons traveling through semiconductors no longer conform to the confines of classical physics. They exhibit wave-like interference, a phenomenon we can harness for efficiency and miniaturization.

Embedded within the transistors of tomorrow are structures so small that quantum mechanics dictate their behavior. The peculiarity of wave-particle duality allows for quantum tunneling – where particles traverse barriers that would, under the Newtonian regime, be impenetrable. By exploiting this principle, semiconductor devices like tunnel diodes are fashioned. They operate on electrons tunneling through barriers, offering faster switching speeds and lower power consumption compared to their classical counterparts.

Now, imagine the precision required to manipulate and harness these subatomic travelers. Enter the realm of nanoelectronics, where materials are manipulated atom by atom, constructing circuits of unimaginable smallness. Here, electrons are treated neither purely as waves nor particles, but as quantum objects governed by a probabilistic rule set. Components like quantum dots operate by confining electrons in three-dimensional traps, manipulating their energy states for LED displays that outshine their forebears with richer colors and deeper blacks.

Adapting to the realities of the quantum world does not cease with active components; it extends to the storage of our digital memories. The dense tapestries of magnetic storage media on our hard drives directly rely on the wave nature of electrons. The recording heads scan the surface, writing data through magnetic fields influenced by the electron's spin, a quantum property tied to wave-particle duality.

Straying away from the silicon jungles and into the observational spyglasses of modern science, microscopy has also been revolutionized by our tiny dualistic particles. Traditional optical microscopes bump against the diffraction limit; they cannot image anything smaller than the wavelength of light used. But in a quantum leap of insight, we've turned to the wave-like nature of electrons. Electron microscopes, wielding beams of these charged particles, enhance our vision to the atomic scale. Through chromatic and spherical aberration correction, the wave nature of electrons is fine-tuned, rendering images with unprecedented resolution.

With an electron microscope, scientists peer into the very building blocks of materials, assessing atom arrangements to unlock the secrets of material properties or plunge into the jungle of biological cells, revealing structures that were once blurred into obscurity. The electron's wave-like diffraction patterns also empower us to perform electron crystallography, deciphering the three-dimensional structure of crystalline materials—imperative for advancements in drug development and materials science.

In the specialized electron microscope variant known as the scanning tunneling microscope (STM), electron tunneling is exploited. By bringing a sharp tip close to a conducting surface but never touching it, electrons quantum leap across the gap. The spatially dependent tunneling current is then mapped, atom by atom, as the tip scans the surface. Such microscopes unmask the atomic landscape of surfaces, granting us the power to observe and even manipulate individual atoms and molecules—a feat that would have seemed sorcery not a century ago.

And what of those moments when we require the simultaneous deployment of particles' wave-like and particle-like traits? The cutting-edge development of holography in electron microscopy does just that. It harnesses interference, a characteristic of waves, to reconstruct three-dimensional images from two-dimensional electron microscope images. In essence, we are encoding and decoding quantum information, a principle that interlaces through the fabric of wave-particle duality.

Stepping back, one marvels at how the conceptual wonderment of wave-particle duality infiltrates our technological veins. It not only underpins the evolution of devices central to our daily lives but also illuminates paths previously obscured at the frontier of knowledge. It's in the embrace of this duality, accepting the bizarre harmony of waves and particles, that our technological symphony finds its exquisite melodies.

So let us not skirt around the edges of perplexity that quantum mechanics presents. Rather, embrace it, delve into it, for it is in the heart of its mysterious duality that our greatest technological advancements lie. From the intricate dance of transistors within our smartphones to the atomic-scale snapshots of life unfurling in the clandestine tapestries beneath microscope lenses—these are testaments to the untold power distilled from the wave-particle duality.

As we continue to explore, innovate, and forge our collective futures, wave-particle duality remains not just a cornerstone of physics, but a beacon that guides us through the quantum age. It propels us into strata of precision and understanding that rewrite the boundaries of what we know and what we can achieve. And in this teeming nexus of abstract quantum principles and their applications, we find the genius of human creativity, in harmonious duet with the fundamental laws that govern the cosmos.

2.2 IMPLICATIONS OF WAVE-PARTICLE DUALITY

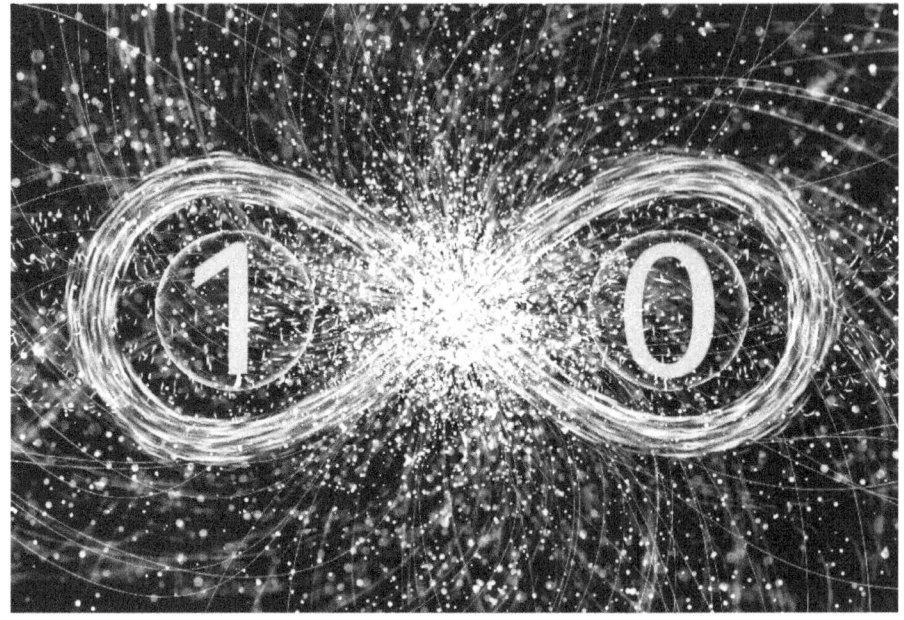

UNDERSTANDING INTERFERENCE PATTERNS

Imagine holding a handful of sand, each grain a different world of possibilities. This is the mesmerizing realm of quantum superposition—a principle so foundational, yet so bewildering, it shatters our classical conceptions of reality. At the core of this chapter lies this extravagant idea: objects can exist in multiple states at the same time until we measure them.

Let me introduce you to one of the most iconic and often misinterpreted thought experiments in quantum mechanics: Schrödinger's Cat. Ponder on a cat sealed in a box, alongside a flask of poison and a quantum mechanism that has an equal probability of killing the cat or leaving it alive. According to quantum mechanics, until the box is opened and the cat observed, it exists in a state of being simultaneously alive and dead. It is our act of observation that forces nature to 'collapse' into one possibility.

At the heart of this paradox is the concept of superposition. Each quantum state—a possible outcome—is akin to a musical note, and a quantum system in superposition is a symphony of these notes. The notes play together, creating a complex harmony of existence that defies the 'one note at a time' ordinary reality. It's a coherent blend of possibilities that is unfathomable in the macroscopic world we live in, yet it is the very basis for the microscopic tapestry of our universe.

But how do we know superposition is real and not just a trick of theoretical mathematics? Enter the celebrated double-slit experiment. When particles such as electrons are fired at a barrier with two slits, they don't simply pass through as individual particles; they interfere with themselves like waves, creating an interference pattern on the other side. This behavior is only possible if the particles travel through both slits simultaneously—in a superposition. The revelation is profound: the particles decide their trajectory only upon being observed. Their journey prior to that moment is a spread-out wave of possibilities.

So, what does it mean for the particle's wave function—the mathematical expression of all possible states—to collapse? It is an instantaneous reduction of all that could be into one definitive state, triggered by measurement. This moment of 'truth' is enshrouded in mystery because it's more than just observing; it's an interplay between consciousness and matter. Each time we measure, we're casting the dice of the universe, forcing its many-faced die to settle on a single number.

A handful of experiments breathe life into this abstract concept. Take the quantum interference experiment, where particles accumulate over time to form an interference pattern—a collective testament to their superposed states. Or quantum cryptography, where particles in superposition enable the most secure communications, exploiting the philosophy that observing changes the observed.

Now let's face a curious observation: quantum systems seem to 'know' when they're being watched an oddity known as the Quantum Zeno Effect. Picture yourself frequently checking on a simmering pot, each peek preventing it from ever reaching a boil. In a quantum system, it appears that constant observation keeps it from evolving, frozen in a speculative stasis of what could be. This startling behavior has implications not only for how we understand quantum mechanics but also for how we can harness its peculiarities for technologies like quantum computers where controlling quantum states matters profoundly.

Nestled in these phenomena is the lingering question: Why does measurement create such turmoil in the quantum world? This is the "measurement problem," an unresolved challenge that quantum physicists continue to grapple with. Some suggest there is no collapse, but rather an unceasing branching of universes (the Many-Worlds interpretation), while others hint at a deep connection between consciousness and the fabric of reality.

Regardless of the theoretical stance, practical implications abound. The applications of superposition are vast and varied, from quantum computing, which uses qubits in superposition to perform complex calculations that would stump classical computers, to quantum metrology, refining measurements to degrees of precision we've never before achieved.

Despite the strides we've made, limitations loom large. The ballet of quantum states is delicate, and any interaction with the environment—a process called decoherence—can disrupt superposition. This is why quantum computers are such finicky creations, requiring extreme environments like near-absolute zero temperatures to maintain their quantum dance.

But do not be discouraged by the enormity of these concepts. Like mastering an instrument, grasping quantum mechanics takes time, patience, and practice. By embracing these quantum oddities, by looking beyond the edge of what we perceive as possible, we unlock a new view of reality. It's a world where a particle can traverse every path in the forest simultaneously, where looking at a star-lit sky could unfold as many universes as there are stars, and where the future of technology rests upon our ability to befriend the quantum world's elusive charm.

Throughout this exploration, remember that you are part of this quantum story. Every interaction you have—each touch, each breath—is a tapestry of countless quantum events. The implausible nature of this reality does not diminish the wonder; it magnifies it, offering a glimpse into the infinite brush strokes that paint our universe's grand mural.

As we part ways with this chapter, let your curiosity linger on the universe's beautiful symphony of superpositions. Let the principles of quantum mechanics resonate within you as both an intellectual pursuit and a metaphysical reflection of our world—a world that's vaster, more interconnected, and more magical than we ever imagined.

Quantum Tunneling and Barrier Penetration

Imagine trying to listen to two different songs at the same time. Each song has its unique melody and rhythm, and yet, if you could hear them both without them interfering with each other, you could potentially appreciate both in a newfound harmony. This, faintly, mimics the heart of **Quantum Superposition** – a concept in quantum mechanics that allows particles to be in multiple states concurrently, challenging our everyday experiences.

In the classical world, we're accustomed to things existing in one state at a time. A coin, for instance, is either heads-up or tails-up when it lands. But suppose we give this coin a quantum twist. Upon tossing this "quantum coin," it whirls in a state that is neither heads nor tails but both – an ambiguous blend. It's not until we catch it, observing it, that the coin 'decides' its orientation. This is the essence of what we term superposition: the coin embodies potential outcomes simultaneously, and observation manifests a single reality.

This isn't just a party trick – superposition is fundamental, casting long shadows across technology fields, including quantum computing, where bits are now qubits, existing in multiple states, ready to perform computational miracles that render today's supercomputers obsolete. Yet, it's not just about faster calculations. It's about different rules for processing information.

But you might ask, "How do we witness superposition if the act of observing collapses these multiple possibilities into one?" Fascinatingly, it's through another mind-bending phenomenon: **Quantum Interference**. When particles like electrons are offered two paths, they don't just pick one; they explore both, interfering with themselves in a way that creates a pattern of highs and lows, akin to ripples in a pond converging and diverging. The pattern exists because of the simultaneous existence of all possible paths, which we can measure.

Enter **Schrödinger's Cat**, a thought experiment you've likely heard referenced in pop culture, which brings superposition into a relatable scale. A cat, a flask of poison, and a radioactive atom are sealed in a box. If the atom decays, the poison is released, and the cat meets its demise. But quantum mechanics says the atom simultaneously has decayed and not decayed, thereby rendering the cat both alive and dead until the box is opened. It's a parable to illustrate the seemingly absurd, yet experimentally supported, world of the very small.

How does this connect with your day-to-day? Let's consider a tech buzzword: **Quantum Computing**. Conventional computers are built on binary, grounded in the certainties of **0**s and **1**s. Quantum computing, on the other hand, harnesses superposition to allow its qubits to be **0**, **1**, or both. Imagine if you could ponder every possible answer to a problem simultaneously. That's the tantalizing promise of quantum computing, promising to revolutionize everything from drug discovery to data encryption.

Yet, there's a catch: the delicate state of superposition is susceptible to **Decoherence**. Interactions with the environment, for example, can 'nudge' the system, causing the superposition to collapse unintentionally. It's the quantum equivalent of trying to whisper in a hurricane – the information you're trying to protect gets swept away by the slightest disturbance. Decoherence is among the premier challenges in developing stable, functional quantum technologies.

The phenomenon that underscores superposition and its collapse is not just a theoretical delight. The **Quantum Zeno Effect** is a bizarre situation where rapidly observing a quantum system seems to freeze it in time, thwarting its evolution – like a watched pot that never boils, but on a subatomic level. It seems counterintuitive, but it's another piece of evidence that observation is not a passive act in the quantum realm.

Yet, it's not all about ephemeral, ethereal states. The practical implications and limitations are concrete challenges in the journey toward leveraging quantum mechanics. Consider the implications for **Materials Science**, where understanding quantum states at the nanoscale can lead to the development of materials with tailored electronic properties – maybe superconductors that operate at room temperature, or solar cells with unprecedented efficiency.

Bang at the center of this quest for practical quantum applications is the development of **Quantum Metrology** – the art of measurement taken to quantum levels of precision. This could mean navigation systems that function without GPS by measuring the Earth's magnetic field with quantum accuracy or medical diagnostics that can sense the magnetic fields of the human brain with such finesse that diseases are detected sooner.

But what do we make of all this? As we pull back and examine the grand landscape of quantum mechanics, these phenomena aren't just quirks of nature to be cataloged and marveled at; they are doorways to innovation, beckoning us to step through. The applications of quantum mechanics in our everyday life may often be invisible, discreetly powering technology, but they are omnipresent – much like the air we breathe. We may not always see it, but its effects are undeniable.

Now let's circle back to our "quantum coin" tossed high into the air, spinning with all the possibility that superposition allows. As it hangs mid-toss, existent in both states, this is quantum mechanics speaking the language of potential – potential not just for the many states a particle could take, but for the future of technology, science, and the fabric of reality itself. And as it descends, we realize that we, as observers, are an integral part of this cosmic conversation, playing a role in shaping the narrative of reality.

As you navigate through the remainder of this book, consider each quantum principle not just as an abstraction or a puzzle to solve, but as an invitation to participate in the unveiling of a universe that is far more intricate and fascinating than we could have ever conceived with just our senses. Keep this thought close: quantum mechanics, complex as it may be, is a story of interconnection, a tapestry where every part informs the whole. In the quest to understand it, we're not just learning about particles and probabilities – we are, in a very real sense, learning about ourselves and our place in the grand scheme of things. Welcome to the quantum world – let the exploration continue.

APPLICATIONS IN ELECTRONICS AND MICROSCOPY

Have you ever found yourself at a crossroads, agonizing over which path to choose, only to wish you could travel both simultaneously? In the captivating world of quantum mechanics, particles lead a life of such possibilities. This is the domain of quantum states and superposition—the heart of quantum mechanics and a phenomenon that shakes our classical worldview to its core.

Understanding Quantum States and Superposition

At its core, a quantum state is akin to a recipe card for a particle. This card divulges all there is to know about the particle, much like how a recipe gives you the rundown of your next culinary adventure. Now, you're likely familiar with binary conditions such as a light switch being either on or off. In the quantum realm, however, things take on a rather spellbinding twist: points between 'on' and 'off' become a canvas of probabilities – a fuzzy in-betweenness where a particle's existence is smeared across multiple possibilities.

Superposition, a term that is as enigmatic as it sounds, is our way of explaining this quantum behavior. Imagine you have a coin. Conventionally, it lands heads or tails when flipped and allowed to settle on your palm. But what if, before it lands, while it is still spinning in the air, I tell you that the coin is both heads and tails? In the quantum world, that isn't just a hypothetical situation; that's superposition. Particles can be in multiple states at once—like our spinning coin, they can be this and that, here and there, until something, say, observes them or interacts with them in some way.

Think back to the Schrödinger's Cat thought experiment we mentioned at the start of this chapter. The cat inside the box is simultaneously alive and dead until you peek inside. The act of opening the box and observing the cat 'decides' the animal's fate. This is the essence of superposition, coupled with measurement, which has astounding implications and applications, as we'll explore shortly.

In the classical world, things are deterministic. If Bill Gates dropped a hundred-dollar bill from the top of Microsoft headquarters, we could predict to a T where it would land, accounting for wind speeds and the aerodynamics of the currency. In the quantum microverse, predictability sails out of the window, replaced by probabilities, giving the quantum state its characteristic representation – the wave function. This wave function is a mathematical expression that encodes all possible outcomes of a particle's state.

Why 'wave' function, you may ask? Because it showcases a profound characteristic – all particles have a wave-like nature, spreading out like ripples in a pond rather than being pinpointed spots. When we measure these ripples, we suddenly find our particle as a discrete entity—a single spot on the pond. This transition from a probability cloud to an actuality upon observation or measurement is known as the 'wave function collapse.'

Now, you might reckon this sounds far out – stuff best suited for philosophers or abstract theoretical physicists. But this peculiar quantum behavior is at the core of modern technology. In computer science, quantum superposition opens the door to quantum bits or 'qubits,' which, unlike classical bits, can be **0**, **1**, or both at the same time. Imagine the processing power when your computer can run through multiple operations simultaneously. It's as if you could read all the books in a bookstore at the same time and comprehend each one perfectly.

But how do we know any of this is real? What evidence do we have to confirm the presence of superposition? The answers are in the experiments. Take the double-slit experiment, for example, where electrons (or photons) fired at a screen with two slits display an interference pattern, signifying wave-like behavior. Yet, if observed at which slit, they pass through, they behave like particles, eliminating the interference pattern. The conclusion is staggering - these particles are in a superposition of passing through both slits until measured.

The implications of quantum superposition cannot be overstated. They ripple through the very foundation of quantum mechanics, rendering it a breathtakingly peculiar yet wondrous landscape, radically different from our everyday experiences. When we dare to venture deep into the quantum world, we encounter phenomena that challenge our notions of reality. It shows the interconnectedness of the universe at a fundamental level that is profoundly more complex and enchanting than what our senses acclimate us to.

Naturally, these insights raise more questions than they answer. Does the universe 'split' with each possible outcome, as suggested by the many-worlds interpretation? Is the universe fundamentally deterministic, but do we simply lack the information needed to predict it, as hidden variables theories propose? Or does the consciousness of an observer play a decisive role in the behavior of quantum systems? These are questions that quantum physicists ponder and debate fervently.

Understanding quantum states and superposition is like acquiring a new sense - a quantum sense - that enables us to peer into the fabric of reality in ways that we never could before. It requires one to let go of classical intuition and embrace the abstract, embrace the quantum. And by doing so, you enable yourself to comprehend a universe that is whimsical, unpredictable, and magnificently mysterious.

With this refreshing perspective on quantum states and superposition, the next section awaits where we will delve into the significance of measurement and observation in quantum mechanics. How does the act of observing alter the state of a system? What does it mean for a particle's fate to be determined upon our intervention? These ponderings are not just for theoretical musings—they're answers that dictate the way we harness quantum phenomena in groundbreaking ways, from cryptography to computing.

So, let us not be daunted by the strangeness of it all. Instead, let us remember that progress often lies beyond the frontiers of conventional wisdom, in the quantum wilderness where possibilities know no bounds.

CHAPTER 3: QUANTUM SUPERPOSITION AND MEASUREMENT

3.1 Concept of Superposition: Schrödinger's Cat Thought Experiment

UNDERSTANDING QUANTUM STATES AND SUPERPOSITION

As we embark on the voyage into the heart of quantum mechanics, you and I have reached a pivotal juncture: the enigma of measurement. Measurement is the process by which we, as observers, pluck a quaint quantum system from a swath of possibilities into the stark light of reality. And here, it's not just about what we measure – it's about how the very act of measuring reshapes the landscape of what is real. Let's delve into this fascinating interplay between observers, measurement devices, and the fragile world of quantum states.

In the classical world, we take it for granted that observing a phenomenon is as innocuous as turning our gaze towards the stars, leaving them unperturbed in their celestial dance. We expect an apple to plummet from a tree identically, whether or not we watch its fall. Alas, in the quantum realm, the act of watching is a far more intrusive affair.

The concept that the observer affects the observed is hardly a new philosophical riddle, but in quantum mechanics, it's not merely thought-provoking—it's an experimental reality. To observe a quantum system, we must interact with it, and that interaction irreversibly changes the system's state—the so-called 'observer effect'.

Picture a single electron journeying through the universe with its bundle of potential locations—a quantum superposition. The second we decide to pinpoint its position, we interact with that electron, and voila, it graciously selects a single spot to occupy. But was our electron there all along, or did the act of measuring compel it to decide? This is the question that tears the fabric of our classical intuitions.

How does this all matter? You may ask. Well, the implications are tremendous. For instance, in the realm of quantum computing, where information is processed in bits that exist in superpositions, the role of measurement dictates not just what we know, but what is knowable. Our capacity to measure certain states without disrupting others is a battle of finesse and ingenuity.

Diving deeper into the quantum sea, we encounter an ally in this journey - the measurement device. Not just a passive tool, but an academic consort that bridges the quantum and classical worlds. A quantum state, when measured by such devices, must 'decide' on an outcome. But this is not a simple case of either/or—these devices must be designed to handle the probabilistic nature of quantum mechanics, a tug of war between chance and certainty.

Here lies the perturbing paradox of quantum measures: while our classical intuition demands the universe to possess definite properties before we measure them, the quantum narrative suggests otherwise. A particle's properties are not defined until they are measured. How do we reconcile this?

We must tread carefully around this tangled garden. One approach to understanding the measurement problem is to treat it as a quantum-classical boundary issue. When does the quantum system stop and the classical measurement apparatus begin? Consider the humble photon—in its unmeasured state, it blissfully exists in both horizontal and vertical polarization states. However, once a polarizer greets it, only one state emerges. Thus, the measurement device itself seems to compel the quantum state to a classical existence.

Yet, this is not the be-all and end-all of the observer's tale in quantum mechanics. We must also consider two fascinating phenomena: the Quantum

Zeno Effect and the wave function collapse. The Quantum Zeno Effect, named after the Greek philosopher Zeno of Elea, suggests that a quantum system's evolution can be halted by the act of constant observation—akin to how Zeno's paradoxes hint at the impossibility of motion.

And the collapse? This is the drastic reduction of the wave function—the mathematical embodiment of all possible states of a quantum system—into a single outcome upon measurement. How and why this collapse occurs is a dance of theories ranging from the practical to the philosophical.

The implications? A mastery of these nuances can revolutionize our world. By understanding the role of measurement, we can unlock doors to unprecedented computational might, harnessing quantum systems to perform tasks unimaginable to classical computers.

However, let us not shy away from the specters lurking in the shadows. The act of measurement introduces noise and decoherence, the sworn enemies of quantum fidelity. We strive to eke out pristine states from a universe that seems hell-bent on scrambling quantum information. What's more, as we push forth into this quantum frontier, we grapple with ethical and societal questions. When does observation breach into surveillance? How do we innovate without infringing upon the myriad fabrics of privacy?

Within our exploration of this quantum measurement tapestry, we take not just a scientific leap but also a philosophical bound. We are not merely passive onlookers in the cosmos; we are participants shaping the very reality we seek to understand. And though our tools may be imperfect and our contexts ever-evolving, our pursuit is clear—to translate the abstract whispers of the quantum universe into a chorus of knowledge that resonates through every fiber of our existence.

In this breeze of contemplation, we alight upon a truth: quantum measurement transcends the mere interaction of device and system. It is a dialogue between the universe and ourselves, each observation a sonnet to the symphony of the cosmos. As we harness this knowledge, not only do we open avenues in technology and innovation, but we also embark on an introspective odyssey—a quest to fathom the depths of reality's fabric and our place within its intricate weave.

Thus, dear reader, we close the chapter on measurement as we open our minds to the boundless possibilities. Our foray into quantum mechanics is not simply an academic indulgence but a journey towards redefining what it means to perceive, to know, and to be. So let us continue, with curiosity as our compass, towards the vast quantum horizon.

EXPERIMENTAL EVIDENCE FOR SUPERPOSITION

In our journey through the rabbit hole of quantum mechanics, we've already danced with the quirks of superposition—how a quantum system can exist in multiple states simultaneously until it's measured. But what happens if we decide to watch the pot to see if it ever boils? This is where the story takes an even more fascinating twist. Welcome to the stage, the Quantum Zeno Effect—a curious phenomenon as enigmatic as its ancient namesake, Zeno of Elea.

Picture yourself incessantly peeking into Schrödinger's infamous box, trying to catch the cat in a single, defined state: alive or dead. The act of observation, in quantum terms, is no mere spectator sport—it's an active player. It turns out, according to the Quantum Zeno Effect, that a quantum system which is observed continuously may be frozen in its state. It's as though your relentless gaze is holding the particles in a quantum traffic jam, preventing them from evolving to any other state.

This concept stems from the very heart of quantum mechanics. Recall our previous discussions on wave function collapse—where the act of measuring snaps a quantum system out of superposition into a definitive state. This collapse is where the Zeno effect draws its juice. Repeated measurements, what you might call 'observation overkill,' essentially hit the pause button on the quantum system's evolution, making transitions between quantum states less likely or even halting them altogether.

The origins of the Quantum Zeno Effect trace back to **1977**, when physicists Baidnayak Misra and George Sudarshan proposed the theory. They named it after Zeno's arrow paradox. The ancient paradox claims that for an arrow to reach its target, it must first cover half the distance, then half of the remaining distance, and so on, ad infinitum, implying that motion is impossible—a clear defiance of common sense. Analogously, if a system is perpetually observed, it never seems to 'move' to a different state.

But why does this happen? Well, it's all about disturbance. When you measure a quantum system, you are not just passively taking in information. You're poking it with a quantum stick. Even the most delicate of measurements perturbs the system, and in quantum mechanics, to perturb is to influence. Each act of observation 'resets' the system, curbing its natural tendency to shift into another state.

Now, I hear the skeptics among you. "If I glance at my pot, does it never boil? That's not how my kitchen works!" And you're right—in the macroscopic world of pots and kettles, observing doesn't freeze dynamics. Here, quantumness takes over. The Zeno effect is an exclusively quantum realm feature, and it highlights the stark divide between our everyday experiences and the bedazzling behaviors of particles at the quantum level.

The implications of the Quantum Zeno Effect are as staggering as they are subtle. For starters, it provides a novel mechanism to control quantum systems. Imagine being able to 'protect' a quantum state simply by keeping an eye on it—a nifty trick for quantum computing, where maintaining coherence of qubits is as crucial as it is challenging. Preservation of quantum information through observation could be a game-changer, potentially enabling us to sidestep some of the hassles of decoherence.

Furthermore, the effect has practical implications in the realm of particle decay. The stability of certain particles could be enhanced by frequent measurement, influencing their decay rates. This could herald new advancements in the control of nuclear reactions or the safe handling of radioactive materials.

Of course, this doesn't mean we can prolong the life of a particle indefinitely by monitoring it incessantly. In practice, there's a trade-off. While continuous observation can hinder the particle's evolution, it's not a permanent stasis field. And given that no observer can maintain an endless watch (and quantum systems are notoriously shy, recoiling from too much attention) the Zeno effect has its limits.

Now, let's address the elephant in the room—the observational influence. This isn't wizardry or the mystical manifestation of consciousness over matter. It's quantum mechanics laying bare a foundational truth: Observers and systems are inextricably linked. When we measure, we get involved. There's no standing on the sidelines; the act of observation becomes part of the system's evolution. Whether we're discussing which-path information in the double-slit experiment, or poking at qubits in a quantum computer, our actions reshape the landscape.

Let's be clear: This isn't about the observer's mind or awareness affecting physical reality in some spiritual sense. It's about physical interactions that take place during the measurement process, and these interactions are governed by the same physical laws that steer the cosmos.

In conclusion, the Quantum Zeno Effect and the role of observation in quantum mechanics invite us to rethink the nature of reality at its most fundamental level. It blurs the lines between the observer and the observed, challenging us to reconsider the archetype of passivity in scientific examination. As quantum technologies evolve, our mastery of these strange effects will not just illuminate the depths of reality—they might just be the key to harnessing the power of quantum mechanics.

This journey—like the never-ceasing dance of particles observed and unobserved—is an ongoing adventure, shaping the future of science and technology. As we press onward, remember that your participation in this quantum universe is not just that of a bystander. You are, in every sense, an active participant in this quantum symphony, capable of influencing the notes played by the very smallest members of our cosmic ensemble.

QUANTUM INTERFERENCE AND WAVE FUNCTION COLLAPSE

As we tiptoe along the precipice of quantum mechanics, let's take a moment to consider a peculiar ally we've encountered: measurement. In our quest to understand quantum superposition and the curious case of Schrödinger's cat, we've stumbled upon a fundamental question: how does one measure a system that is a chameleon, a master of being in more than one place at once? It's a bit like trying to pinpoint the exact location of a buzzing bee as it darts from flower to flower. The act of observing, it seems, has its own set of implications and limitations in the quantum world.

Measurement, in classical physics, is somewhat of a routine affair. You take a yardstick, measure your garden, and expect the same outcome each time, provided you measure correctly. Quantum mechanics, on the other hand, is the unruly teenager of the physics family – it plays by its own set of probabilistic rules. When we measure a quantum state, we're not just passively looking at a property that's waiting to be discovered; we are actively choosing one outcome out of many possibilities. It's as if by merely glancing at a die, you could force it to settle on a six, every single time.

The Role of the Observer in Quantum Measurement

The notion that the observer can affect the system they're measuring is a wild departure from anything we've seen in the realm of classical physics. While we won't delve into the philosophical implications here, it's important to note that experiment after experiment confirms this strange behavior. Quantum systems remain in a superposition of states until we come along with our measurement devices. The wave function – the mathematical representation of these probabilities – collapses, and a single reality crystallizes from the haze of probabilities.

Quantum Zeno Effect and Measurement Frequency

The Quantum Zeno effect adds another layer to this story. It suggests that the more frequently we take measurements of a quantum system, the less likely it is to evolve. The name is drawn from Zeno's paradoxes of motion; just as Achilles paradoxically never overtakes the tortoise if one keeps dividing the remaining distance between them, a quantum system seems to freeze in place under constant observation. In more concrete terms, this effect could, in theory, prevent radioactive decay simply by observing the atom often enough.

Heisenberg's Uncertainty and the Limits of Knowing

We cannot discuss quantum measurement without bowing to Heisenberg's uncertainty principle. Traditionally introduced with position and momentum – the more precisely we know one, the less we know the other – this principle applies to various paired properties in quantum physics. This isn't a limitation of our instruments or methods; it's a built-in feature of the universe. Nature demands a trade-off, a kind of cosmic quid pro quo, and it is intrinsic to the quantum fabric of reality.

Decoherence: The Environment's Role in Measurement

Our quantum system, as isolated as we try to make it, is never truly alone – it's always influenced to some extent by its environment, leading to decoherence. This process, in which superpositions seem to "degrade" when a system interacts with its surroundings, poses a significant hurdle in quantum computing. Keeping a quantum bit, or qubit, coherent long enough to perform operations before decoherence kicks in, is like trying to whisper a secret in a bustling crowd – the message tends to get lost.

Entanglement: Measuring One, Affecting Another

Entanglement muddies the waters of measurement even more. When two particles are entangled, measuring one instantly affects the other, no matter the distance separating them – a phenomena Einstein famously referred to as "spooky action at a distance." When we measure an entangled particle, we inadvertently collapse the wave function of its distant partner. It's akin to two silent dancers on a stage, separated by a great distance; as soon as one takes a step, the other mirrors it instantly.

The Observer Effect vs. Observer-Independent Reality

It's essential to distinguish between the observer effect, where the act of measurement influences the system, and the idea of observer-independent reality, where things exist and have properties whether we look at them or not. Quantum mechanics blurs this line, and in some interpretations, there's no reality without observation. Wipe that cold sweat from your brow; this doesn't lead to solipsism – the notion that only one's mind is sure to exist. The scientific community continues to grapple with these concepts as they explore the foundations of quantum theory.

Practical Implications for Technology

In practical terms, the nuances of quantum measurement are key in developing technologies like quantum cryptography. A message encoded using quantum key distribution is theoretically impervious to eavesdropping because any attempt at measurement would disturb the system and reveal the intruder. Along similar lines, quantum computers capitalize on measurements to manipulate data in ways classical computers can't fathom.

Limitations in Precision and Control

Even as we dream of the era of quantum superiority, where quantum computers outperform their classical counterparts, we face limitations in our ability to make precise and controlled measurements. Quantum systems are delicate; they don't take kindly to aggressive probing. We face a delicate balancing act – interact with the quantum system enough to harness its advantages but not so much that we destroy the very properties we wish to use.

Addressing the Challenges

Researchers are engaged in a constant battle to refine measurement techniques. Error correction codes and better isolation of quantum systems are among the strategies employed to mitigate the adverse effects of measurement. There's a growing understanding that quantum systems, with all their eccentricities, require a unique blend of finesse and technological prowess to yield their secrets.

In the realm of quantum measurement, we find ourselves at a crossroads of understanding and application. Every attempt to measure nudges the boundary of our knowledge a bit further, yet also reveals the inherent limitations of our reach. It's a stark reminder that when dealing with the quantum realm, each answer we find seems to unlock a further set of mysteries – an ever-expanding puzzle for the endlessly curious and the daring pioneers of science.

3.2 Measurement in Quantum Mechanics

ROLE OF OBSERVERS AND MEASUREMENT DEVICES

As we unfurl the fabric of the quantum realm, we unearth a phenomenon so peculiar and profoundly impactful that it reshapes our understanding of connectivity and information itself: quantum entanglement. Often described through a narrative involving particles behaving like telepathically linked twins, quantum entanglement has long been a subject shrouded in mystery. Yet, it's poised on the cusp of revolutionizing how we communicate, compute, and comprehend the universe.

To grasp the concept of entanglement, we must first appreciate that particles can exist in a state of deep connection regardless of their separation in space. This is not the kind of everyday interaction you might have with your phone or computer; it's an instantaneous, silent dialogue. If you alter the state of one particle, its entangled counterpart reflects this change instantaneously, a phenomenon that Einstein famously dubbed "spooky action at a distance." But there's nothing supernatural about it; quantum entanglement is deeply embedded in the fabric of reality.

Now, you might be wondering how these entangled states are created. It all starts when two or more particles, such as photons or electrons, interact in ways that make their properties, like spin or polarization, dependent on each other. From that point on, even when separated by vast distances, the measurement of one particle's property instantly sets the corresponding property of its partner. This might seem at odds with the snail-paced traffic you battle during your morning commute, but quantum particles play by an entirely different set of rules.

Einstein's reservations are immortalized in the Einstein-Podolsky-Rosen paradox (EPR), an intellectual challenge to quantum mechanics suggesting that the theory was incomplete. While tempting to agree, numerous experiments have vindicated quantum theory, leaving us with a bizarre yet undeniable truth: two entangled entities are so intimately linked that they behave as one unified system, no matter the space between them.

Enter Bell's Theorem. Physicist John Stewart Bell proposed a way to test the reality of these quantum correlations against any theory advocating local realism, which postulates a world where properties are defined prior to and independent from observation, and nothing can influence something else faster than light. Bell's inequality—an elegant mathematical inequality—serves as a litmus test. The results? Nature sides with quantum entanglement, leaving local realism an ideal of classical physics.

These entangled states are fascinating not just for their philosophical implications but for their practical applications. In the world of cryptography, entangled particles are the equivalent of the unbreakable code. Think of it as having an unsnoopable line of communication. If someone tries to intercept your quantum message, the mere act of measuring one particle alters the entire system, revealing the presence of an eavesdropper. This is the cornerstone of quantum cryptography, which promises a level of security that is fundamentally underpinned by the laws of physics rather than computational complexity or crafty codes.

There's more on the entanglement horizon—quantum teleportation. While it's not about beaming humans from one planet to another, it is the transmission of information from one particle to another over arbitrary distances, without the need to traverse the physical space between. It's as if you could send a thought directly into your friend's mind from across the world, bypassing speech, writing, and digital messaging.

Surely, you're thinking, there must be a catch. And indeed, quantum entanglement is delicate; creating and maintaining this entangled state requires incredible control and isolation from environment-induced 'decoherence.' Scientists and engineers are grappling with these challenges, devising clever quantum systems to exploit these phenomena.

While the dawn of quantum technologies seems like a chapter from a science fiction novel, we're closer than you might imagine. Companies and researchers are tirelessly weaving the threads of quantum entanglement into the fabric of emerging technologies, promising an era where quantum networks are as commonplace as the internet is today. Can you envisage a world where quantum sensors so precise could diagnose medical conditions before symptoms even arise, or where quantum reasoning reshapes the face of artificial intelligence completely?

These are not just whimsical predictions but the tangible horizon of quantum entanglement applications. The amassing efforts within quantum communication networks signal a leap toward a future rife with previously unthinkable prospects in security, computation, and beyond.

Yet, for all its potential, quantum entanglement doesn't come without its fair share of conceptual head-scratchers and technical obstinacy. The phenomenon flouts the traditional narratives of causality and locality that our common sense is built upon. This is why figures like Einstein struggled with it. However, in the quantum sphere, our intuition must bow to the oddities that experiments unfailingly confirm, opening our eyes to a more interconnected, yet stranger universe than our ancestors could have fathomed.

As we continue to walk the path of understanding and leveraging quantum entanglement, the fusion of theory and application demonstrates the essence of science—turning 'spooky' mysteries into engines of innovation that propel us into the future. The potential of entangled states beckons a reimagining of communication, computation, and even the fundamental grasp of reality; embedding in our collective consciousness the notion that there is so much more to discover when one dares to peek beyond the quantum veil.

In the grand tapestry of reality, each thread of knowledge we pull brings us closer to the realization of the dream woven by the pioneers of quantum physics, a dream where the 'spooky' is demystified and harnessed to transform the very world we inhabit. And as we thread this labyrinth of understanding, each unraveling mystery leaves us poised at the brink of a new and uncharted quantum epoch.

QUANTUM ZENO EFFECT AND OBSERVATIONAL INFLUENCE

Understanding the wonders of quantum mechanics can sometimes feel like trying to solve a puzzle without seeing the big picture. Yet when we start to piece together this mind-bending puzzle, we illuminate aspects of our world - and ourselves - in profoundly intricate and unexpected ways.

Imagine you're sifting through a patch of dirt in your backyard when you stumble upon a strange, shimmering stone. This stone is unlike any you've seen before; it defies the laws of physics as you know them, changing color and shape when observed from different angles, appearing to be in more than one place at once. That stone is a metaphor for quantum mechanics in our everyday lives—a mysterious and pervasive element that transforms our understanding of everything around us.

Now let's ponder on the role and impact of observers and measurement devices in quantum mechanics. It's a bit like that childhood game of peekaboo: the act of observation can dramatically influence the state of quantum systems.

In classical physics, what you see is what you get. Drop a ball, and it falls; nothing more complex than that. This simplicity, however, gets thrown out of the window when we delve into the quantum realm. Here, the act of observing doesn't just reveal the system's state but can actually determine it.

This eerie quantum realm opens its gates when we consider the case of a quantum particle existing in a state of superposition. Picture it like this: a quantum particle can travel down multiple paths simultaneously. It's only when we decide to peek - when a measurement is made - that the particle 'chooses' a specific path. It's akin to having multiple possible futures, and the act of observation selects which future comes to pass.

But what exactly is an 'observer'? You might think it's you or me with a magnifying glass or some sophisticated apparatus. Although humans can fulfill this role, in quantum theory, an 'observer' can be anything that interacts with the quantum system and forces it to adopt a definite state. It might be as simple as a photon bouncing off an atom or a stray particle from the cosmic background radiation.

One of the most fascinating elements of this quantum story is the Quantum Zeno Effect. This is named after the Greek philosopher Zeno of Elea, known for his paradoxes about motion and change. In quantum mechanics, this effect is the peculiar phenomenon where a system's evolution can be halted by frequent measurements. You can think of it as repeatedly peeking at the quantum system, essentially freezing it in time like a video game character paused mid-action.

To wrap your head around this concept, imagine watching a pot of water waiting for it to boil. In our everyday experiences, watching the pot doesn't prevent it from boiling. However, in the quantum world, consistently observing a quantum system can prevent it from changing state. It's like the quantum pot of water never boils because you're watching it too closely.

It's worth noting that these bewildering quantum phenomena have practical implications, setting the stage for revolutionary technologies. For instance, the measurement process is central to quantum computing, where a qubit's state is determined when measured. Moreover, the awareness of our influence on quantum systems is vital in the development of precise measurement tools, such as in the field of quantum metrology.

Yet, all of this wonder and peculiarity does come with limitations. Each measurement disturbs the quantum system to some extent, potentially altering the information we seek to obtain. This introduces 'noise,' akin to static on your radio that obscures the music. In quantum experiments, this noise can cloud the picture, making it tricky to discern the true nature of a system.

This limitation connects to a broader question which haunts quantum mechanics: Is there a line defining where the quantum world ends and the classical world begins, or are our observations merely a window into a part of the universe we don't yet fully comprehend? These are not just philosophical musings; they're practical questions that researchers grapple with as they design experiments and technologies.

As we seek to apply quantum mechanics in the real world, the challenge lies in handling quantum systems without destroying their delicate quantum state.

We need to strike a careful balance—observing enough to harness quantum properties while not observing so much that we wreck the quantum magic.

Think of a tightrope walker, gracefully poised midway above a vast chasm. On one side lies too much interference, toppling the quantum properties we depend on. On the other side, seeing too little risks stepping into the abyss of uncertainty, with no useful information. The quantum researcher's job is akin to guiding the walker to reach the desired platform, benefiting from the full potential of quantum mechanics without tumbling down.

As we come closer to the end of our quantum journey, take a moment to recognize the extraordinary implications of these principles. By refining how and when we observe quantum systems, we're not just probing the fundamentals of reality—we're forging new tools that will drive forward technology and our comprehension of the universe. From the components in your smartphone to the fabric of spacetime itself, the subtle dance of quantum mechanics and the act of observation is everywhere.

It's all about balance—knowing when to watch and when to look away. As quantum mechanics integrates further into the scaffold of our daily lives and technology, our role as observers in the universe becomes increasingly significant. Who knows how deep the rabbit hole goes when we start to question not just what we see but the very act of seeing itself? That's the quantum world—endlessly fascinating, inherently mysterious, and central to the ever-evolving narrative of human understanding.

PRACTICAL IMPLICATIONS AND LIMITATIONS

Imagine a pair of star-crossed lovers parted by vast distances, yet somehow, when one feels joy or despair, the other, no matter how far away, reacts as if the feelings were their own. This romantic analogy befits one of the most enthralling phenomena of quantum mechanics: quantum entanglement.

Entanglement begins at the microcosmic scale, involving particles such as electrons or photons. When two particles become entangled, they form a connection that is independent of the space between them. Change the state of one particle, and the other changes instantaneously—it's as if they share an invisible thread that ties their fates together.

To understand the characteristics of entangled states, one should first appreciate that in the world of the very small, particles are represented by wave functions—a mathematical expression of the particle's state. This wave function encapsulates the probability of finding a particle in particular places or with particular properties.

When particles become entangled, their individual wave functions meld into a single wave function, encompassing both particles' probabilities. This combined wave function needs to be considered as a whole, and no longer can you describe one particle without considering the other, no matter the distance between them—a monumental break from the separable, individual existence to which we're accustomed.

This entangled state throws a wrench in our classical understanding of the world. Prevailing wisdom suggests that for one object to influence another, some force or signal must traverse the space separating the two, limited at most by the speed of light. Entanglement sidesteps these limitations, operating outside the purview of classical causality. When an observable property known as a quantum state, say the spin of an electron, is measured in one of the entangled partners, the other seems to 'know' about this measurement and exhibits a complementary spin immediately.

The remarkable nature of this phenomenon elicited the famous phrase from Einstein, who referred to it as "spooky action at a distance." His discomfort sprang from the implication that information between entangled particles appeared to be transmitted instantaneously, surpassing the speed of light, which stands in defiance of his theory of relativity.

Einstein wasn't alone in his skepticism. Many physicists wrestled with the implications of quantum mechanics, attempting to reconcile them with the deterministic world of classical physics. However, as experiments advanced, they consistently supported the quantum view. Entanglement was not a quirk of incomplete theory or flawed experimentation; it was a fundamental feature of our universe.

Another gripping aspect of entanglement is that it is indiscriminate of the kind of particles involved. Atoms, photons, and even larger molecules can entangle, suggesting a pervading role for entanglement in the fabric of reality.

Moreover, the uniqueness of entanglement doesn't end with mere connection. The quality of entanglement, or entanglement 'strength,' can differ. Scientists measure this quality in terms of 'entanglement entropy'—a higher entropy denotes a stronger, more complex entangled state. This opens up a landscape of possibilities for how entangled particles can interact and the complexity of states they can exhibit.

Furthermore, entanglement has a property called 'monogamy,' which means that if two particles are maximally entangled with each other, they cannot be entangled, even the slightest, with anything else. It's a fidelity in the quantum world that makes entangled particles cringe at the idea of a crowded relationship.

However, entanglement is a delicate state. The moment an entangled particle is measured or interacts too strongly with its surroundings, its wave function collapses and the entangled state disintegrates. This is known as 'decoherence,' the bane of every physicist working to harness entanglement for practical use. The ability to control and protect entangled particles from external influences is a monumental challenge in quantum technology.

The understanding of entanglement and its characteristics has profound implications, not just for fundamental science, but for the technological marvels they could enable. Quantum cryptography, for instance, uses entanglement to establish secure communication channels where eavesdropping can be instantaneously detected, thanks to the fragile nature of entangled states. There's even the wild frontier of quantum teleportation, where the properties of one particle can be 'transported' to another, thanks to their entangled link.

Science's march forward has only deepened our understanding and appreciation for this quantum oddity. Experiments conducted over increasing distances on Earth, and even via satellites, continue to affirm entanglement's defiance of classical explanations. However, it remains true to the principles outlined by quantum mechanics—an indelible link from the quantum realm to our macroscopic world.

The exploration of entangled states is not merely an academic exercise. It is the pathway to a new paradigm of technology and understanding of reality itself. Each experiment, each theoretical advancement, draws back the curtain a little more on this profound connection, revealing the quantum threads that may yet weave the fabric of the universe.

CHAPTER 4: QUANTUM ENTANGLEMENT AND SPOOKY ACTION AT A DISTANCE

4.1 Introduction to Quantum Entanglement

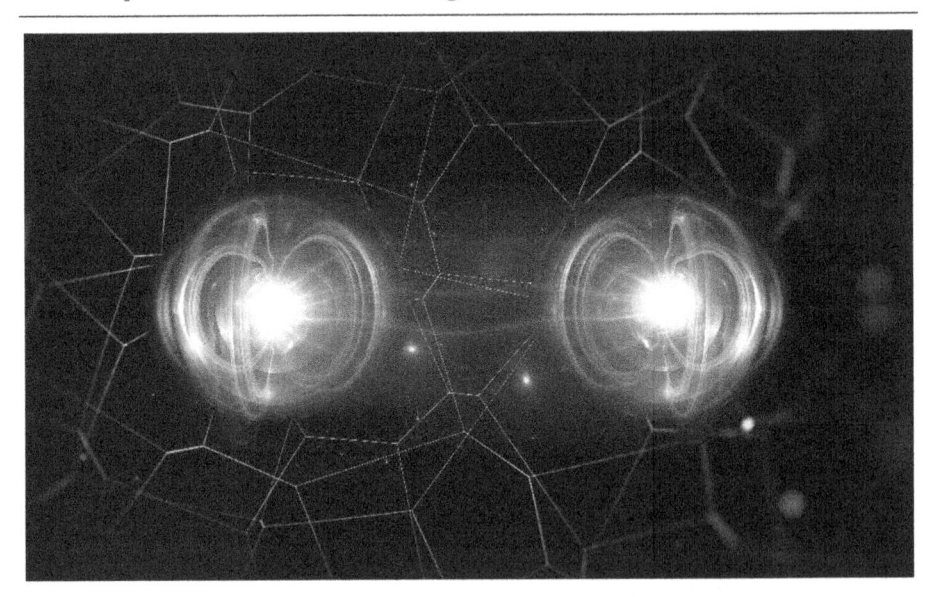

Definition and Characteristics of Entangled States

In the mesmerizing dance of quantum entanglement, where two particles perform a mysterious tango irrespective of the distance separating them, there lies a blueprint for revolutionizing the way we communicate. At the heart of this revolution is quantum cryptography—the utilization of quantum properties to forge a new frontier of secure communication. But how does this seemingly magical system work, and what does it portend for the future of information transfer?

Quantum cryptography owes its security to the fundamental principles of quantum mechanics. One of its most well-known protocols, Quantum Key Distribution (QKD), utilizes entangled photons—particles of light—to share encryption keys between parties. Unlike classical encryption that can be hacked with enough time and computational power, the encryption provided by QKD is underpinned by the laws of physics, making it theoretically unbreakable.

Picture this: Alice and Bob are two communicators who wish to exchange a private message. To create a secure channel, they first need to share an encryption key—a series of bits used to encode and decode the message—without the risk of interception by an eavesdropper, whom we'll call Eve. In the quantum world, Alice generates a sequence of entangled photon pairs, sending one photon of each pair to Bob, whilst retaining the other. Each photon has a polarization state that can represent a bit of information—a **0** or a **1**—for the key.

Here's where the enchantment of quantum mechanics comes into play. If Eve attempts to spy on the transmission by measuring the polarizations of the photons sent to Bob, the act of measuring disturbs the entangled state due to the observer effect. This disturbance can be detected by Alice and Bob, revealing Eve's presence. They can then discard the compromised key and start anew. This process, my friends, is known as the "no-cloning theorem" of quantum theory, which forbids the duplication of an unknown quantum state and thus ensures the security of quantum keys.

The elegance of QKD doesn't stop at its defense against eavesdropping—it also offers forward secrecy. Even if Eve collects encrypted messages, without the quantum key, she can't decrypt them. Not today, not tomorrow, not even with the most advanced future quantum computer.

But it's not just about securing the key; the transmission of quantum information also lends itself to another mind-bending application: quantum teleportation. Although teleporting humans across space à la science fiction remains elusive, we have nonetheless mastered the art of teleporting information through the phenomenon of entanglement. Think of it as a quantum email, where the information encoded in a particle's quantum state can be instantaneously transmitted to a distant entangled partner. This doesn't violate the speed of light—the "no teleportation theorem" ensures that; it's a subtle shift in state, not an actual particle zipping across space at superluminal speeds.

Lest we get carried away by enthusiasm, it's important to note the challenges that lie ahead. Implementing these sophisticated quantum communication networks necessitates advanced technology and infrastructure. The delicate quantum states of photons can be easily disrupted by environmental factors like noise and loss of signal over long distances in optical fibers or through the air, which are current hurdles for practical QKD systems. Furthermore, QKD requires authenticating the communicators to prevent a third party from pretending to be one of the legitimate users.

But herein lies the excitement—the quantum realm is not a static exhibit but a laboratory of continuous innovation. Researchers are pioneering new protocols to extend the range and robustness of quantum communications. Quantum repeaters, much like the repeaters in classical communication networks, could bridge greater distances, while satellite-based QKD has the potential to weave a global network of quantum exchanges.

The implications of quantum secure communication stretch far beyond encrypted emails and secure transactions; it lays a framework for a potential quantum internet. This future quantum network could connect quantum computers, sensors, and other quantum devices, allowing for levels of data security, computational power, and communication efficiency we've only begun to imagine.

As we peer into this quantum horizon, it's clear that challenges await—both technical and ethical. Quantum cryptography raises questions about surveillance, privacy, and the balance of power. With data breaches and digital espionage ever-present in our headlines, the demand for ironclad security is unrelenting.

In this quantum age, we stand at the cusp of a paradigm shift in the fidelity of communication. A world where secrets remain secret, where private conversations ripple through the quantum foam untouched by prying eyes, is within our grasp. As daunting as the journey may appear, our pursuit for understanding and mastery of the quantum world fuels an inextinguishable hope—a hope for a future where quantum cryptography and secure communication are not just novelties but normative standards in safeguarding our digital lives. This is the exciting quantum leap awaiting us just beyond the event horizon, and together, through persistence, curiosity, and the indomitable human spirit, we will navigate this quantum tide.

BELL'S THEOREM AND VIOLATION OF LOCAL REALISM

Imagine you are at a grand magician's show, and the magician performs the classic 'Teleportation' illusion, where an object vanishes from one place and reappears instantaneously in another. Now, suspend your disbelief for a moment, because in the quantum realm, this is not an illusion but a fascinating reality known as *quantum teleportation*. It's a process so mind-bending that it could only come from the pages of science fiction, yet it is a thriving area of research in fundamental physics and a potential game-changer in how we transmit information.

In the quantum universe, information can be transported from one location to another without traversing the physical space between them. This doesn't mean we can teleport people or objects à la Star Trek—sorry to disappoint the Trekkies amongst us. Instead, quantum teleportation involves the transfer of *quantum information*—specifically, the exact state of a particle—to another particle some distance away. This is possible thanks to the peculiar properties of quantum entanglement, which we've explored earlier in this book.

The Quantum Magic Trick Explained

To teleport quantum information, you need a pair of entangled particles. These two particles are in a special state where the actions performed on one affect the other, regardless of the distance separating them. Let's label these entangled particles as Alice and Bob—a nod to the placeholders often used in scientific thought experiments. Now imagine a third particle, Charlie, which holds the information we want to teleport.

The process begins when Charlie's quantum state is transferred onto Alice through a delicate operation called a *Bell State Measurement*, intertwining Charlie's properties with Alice's. The conundrum here is, Charlie's state is actually destroyed in the process—a nod to the no-cloning theorem of quantum mechanics, which forbids the creation of an identical copy of an unknown quantum state.

Now for the magical part. The outcome of this measurement is then classically communicated to Bob's location. I say "classically" because this part uses regular, non-quantum means, such as a telephone call or an email. Once Bob's keepers receive this information, they perform specific quantum operations on Bob that depend on the outcome of Alice's measurement. Miraculously, Bob then assumes the exact state that Charlie had before the procedure started. Voilà! The state of Charlie has been teleported to Bob.

Thinking Beyond Space

When considering quantum teleportation, it's crucial to understand that it's not about sending an object from one place to another. Instead, it's about transferring the very essence of a quantum state across the expanse of space without the state being transmitted through the intervening space. This means there's no "quantum signal" zipping through space-time; it's a transfer of information borne out of entanglement.

Also, there's another common misconception that quantum teleportation happens instantaneously. While the entanglement part is indeed instantaneous, adhering to Einstein's 'spooky action at a distance', the classical communication required to complete the process is bound by the speed of light. Yes, even quantum mechanics respects the universal speed limit set by relativity.

A Future Wired by Quantum Threads

So, what does quantum teleportation mean for the future of information transfer? Think of a world where the Internet is replaced by a 'Quantum Internet', where information is transmitted with absolute security guaranteed by the laws of physics—no more fears of eavesdropped emails or hacked data transfers. Quantum teleportation could enable us to create quantum networks where information is shared between quantum computers across the globe, revolutionizing computation, communication, and cryptography.

Forging Unhackable Communications

One promising application of quantum teleportation lies in the field of quantum cryptography, particularly in creating secure communication channels. Unlike classical cryptography, which relies on mathematically complex problems to deter hackers, quantum cryptography is protected by the unyielding rules of quantum mechanics. An eavesdropper trying to intercept a quantum communication would inevitably disturb the quantum states involved, making the intrusion immediately detectable. This kind of security is invaluable, especially in an era where digital privacy is continually at risk.

Quantum Computing's Far-reaching Arms

Advances in quantum teleportation feed directly into the emerging realm of quantum computing. As quantum processors become more complex, the need for quantum information to be relayed reliably and without degradation is paramount. Teleportation could provide a means to move quantum information around within a computer or between computers without the loss of coherence or information—a critical challenge that currently impedes the scalability of quantum computing systems.

Facing Challenges Head-on

But all this is not without its challenges. The teleportation process requires a high degree of precision and control over individual particles, a task that is, to say the least, non-trivial. Entangled particles also need to retain their linked properties long enough to perform the necessary operations, something jeopardized by the erratic phenomenon of decoherence, where quantum systems lose their peculiar properties and succumb to the ordinariness of the classical world.

The scalability of quantum teleportation is another hurdle. Teleporting the state of a single particle is impressive, but for quantum networks to flourish, we need to execute teleportation operations on a much larger scale. With each incremental increase in complexity, the challenges grow exponentially, requiring leaps in technological prowess and experimental finesse.

A Quantum Leap Forward

Despite these challenges, progress is being made at a breathtaking pace. Through tireless experimentation and lossless determination, researchers continue to push the boundaries of quantum mechanics, making what was once a theoretical wonder into a practical tool. Quantum teleportation has already been demonstrated over several kilometers of fiber optics and even through open air, signaling the early whispers of a revolution in communication and computing.

Quantum teleportation offers us a glimpse into a future where the fabric of information transfer is woven with the fine threads of quantum mechanics. It challenges our perceptions, tests the limits of our understanding, and promises a world where the exchange of information is bound by the unassailable entanglement of particles across the universe. So let us marvel at the magic of these 'spooky' actions, for they illuminate the path toward a future as wondrous as any trick pulled from a magician's hat.

EPR Paradox: Einstein, Podolsky, and Rosen

As we venture deeper into the maze of entanglement, one may feel as though they are journeying through a mystical forest from an enchanted tale, where every path promises the thrill of discovery. Quantum entanglement, that elusive phenomenon where particles become so interconnected that the state of one instantaneously influences the other, regardless of the distance separating them, captivates and puzzles us in equal measure. Albert Einstein famously dismissed it as "spooky action at a distance," yet today, it stands at the precipice of revolutionizing our world. The challenges and potential it holds are akin to uncovering a hidden language of the universe, one that we are only beginning to comprehend.

One of the most profound hurdles we face in entanglement research is the phenomenon of decoherence. Imagine, if you will, trying to listen to a delicate melody in the midst of a roaring cacophony. The pristine state of entangled particles is swiftly disturbed by their environment, causing them to lose their entangled properties—a challenge akin to preserving the sound of a whisper in a hurricane. This sensitivity to noise imposes stringent conditions on our experiments and potential applications, requiring innovations in isolation and error correction that could be as complex and elaborate as the entangled states themselves.

Yet, where there are challenges, there are also trailblazers relentlessly pushing boundaries. Quantum repeaters, devices that extend the distance over which entanglement can be maintained, are being refined to facilitate long-range quantum communication networks. These networks could transform secure data transmission, offering a level of privacy and security matched only by the fundamental laws of physics. Despite the progress, scaling these systems to a global level is an engineering endeavor so ambitious, it rivals the construction of the internet itself.

The field of quantum cryptography, too, faces the test of practical implementation. Quantum key distribution (QKD) harnesses entanglement to create theoretically unbreakable encryption keys. Converting this theory to a ubiquitous technology is no minor feat, requiring not just advanced technical solutions to extend the distance over which QKD can operate, but also a reimagining of our current infrastructure. The task is equivalent to reweaving the very fabric of our digital communications, where the new thread is a quantum one.

A particularly salient question within entanglement research is how to measure it without destroying it. Entanglement is a delicate flower that withers under the gaze of traditional observation techniques. This sensitivity mandates the development of novel detection methods that can preserve entanglement long enough for it to be harnessed. The refinement of these methods could open the door to quantum teleportation, not of matter, but of information. This technology is still in its embryonic stage, much like replicating a whisper across oceans, yet it promises an epoch where information can be transferred without traversing the physical space in between.

One emerging field birthed from the womb of entanglement research is that of quantum metrology, which seeks to leverage entangled particles to make ultra-precise measurements. These measurements could revolutionize navigation systems, timekeeping, and even our understanding of the fluctuating earth beneath our feet. The challenge lies not only in refining the entangled systems but also in integrating them seamlessly into devices and instruments that can be used outside the quantum laboratory.

Inescapably, entanglement research walks hand-in-hand with ethical and societal implications. As we unwrap the gift of these new technologies, we must also grapple with their impact on privacy, security, and the distribution of power. The potential of a quantum internet underpinned by entanglement is no small wonder, offering an interconnectedness beyond our current capabilities. However, it ushers in concerns about surveillance, data ownership, and the widening chasm between those with access to this technology and those without.

What stirs the imagination is not the technological convergence alone but the profound questions that entanglement poses about the fabric of reality itself. Entanglement challenges our classical intuition of separability and locality. As we peel away layers, we may find answers to the nature of spacetime and the enigma of quantum gravity. Imagine a tapestry where these threads of entanglement weave through the very essence of existence, binding the microscopic quantum world to the vastness of the cosmos—a tapestry that we are only beginning to sketch.

The uncharted future of entanglement research is vibrant and fertile, promising advancements in our ability to manipulate and understand this quintessential quantum resource. As we hone our quantum toolset, we will advance closer to the quantum computer, a mighty oracle of computation, which could tackle problems that stump classical machines. The roadmaps laid out by physicists and engineers will guide us through a labyrinth where every turn could reveal a new secret or an unexpected obstacle.

Finally, we must cultivate a quantum-literate society that can reap the fruits of this quantum revolution. Efforts to demystify and disseminate knowledge about entanglement and its applications must be accelerated, as public appreciation and support will be as crucial as the scientific breakthroughs themselves. Imagine a world where conversations about entanglement and its implications are as common as those about the latest smartphone or the Internet of Things. It's a future where quantum literacy is not just for the scientific elite but a shared language for humanity as we step boldly into this strange, quantum-entangled world.

As we look to the horizon, the promise of quantum entanglement is a testament to the indefatigable human spirit—the same spirit that once dreamt of flight and voyages to the moon. Entanglement research embodies the pinnacle of our quest to unveil nature's secrets: a dance between the graspable and the yet-to-be conceived, a thoroughfare to an era where the quantum and the classic realms meld in ways once relegated to the realm of science fiction. Join me, then, as we leap into the quantum future, a realm vibrant with possibility, and let's turn the 'spooky action' into the most profound action of our time.

4.2 Applications of Quantum Entanglement

Quantum Cryptography and Secure Communication

In the beguiling world of quantum mechanics, one of the most profound phenomena we grapple with is the concept of quantum superposition. At the crux of this idea is the suggestion that a quantum system can exist in multiple states simultaneously—until it doesn't. Let's sail together into the paradoxical sea where a single quantum particle can be at different places at the same time and unravel what it means when we talk about reality.

Imagine walking into an art gallery where a painting simultaneously displays a sprawling ocean and a bustling cityscape. That is the kind of duality quantum particles exhibit in their natural habitat. Take electrons, for instance. When not being observed, they seem to explore all possible paths around the nucleus of an atom, living a life filled with an array of possibilities. But once we look closer, trying to pinpoint their exact location, they snap out of this dreamy state and pick a single reality for us to observe.

How does all this manifest in a real-life scenario? Picture the technology in your smartphone. At its heart, you'll find a semiconductor chip that is essentially basking in the exotic dance of electrons living in a haze of superpositions, enabling the rapid computation and processing that powers your device.

The most illustrious parable for quantum superposition is the Schrödinger's cat thought experiment. A cat, a flask of poison, and a radioactive source are placed in a sealed box. If a single atom of the substance decays, the poison is released, and the cat meets its demise. The quantum kicker? Until that box is opened and the result observed, the cat is considered to be both alive and dead. It's a mind-bender, to say the least.

But beyond thought experiments, superposition is an observable fact. Take the example of quantum interference. Sometimes, particles such as electrons or photons are fired one at a time at a double-slit apparatus, which results in a pattern on the other side that implies that each particle interfered with itself. This can only occur if the particle was in a superposition of passing through both slits at once, before being measured at the screen.

Measure once, you might catch our particle basking left of center; measure again, it might be lounging on the right. Just like asking different witnesses for their account of an event, the act of measurement in the quantum realm retrieves various tales from particle states, yet shrouded in a mist of probability.

The quantum interference pattern is evidence of this phenomenon—each particle riding the waves of possibility, creating a tapestry of light and shadow that could only result from a dance of existence in many places at once. And with the collapse of the wave function, a story of what could have been is pruned to the narrative of what is—one reality out of many.

Yet, this isn't mere quantum navel-gazing. Superposition is the cornerstone of technologies that are fast changing the world. Quantum computers, for example, harness this principle to perform operations on a scale unfathomable to classical computers. In these über-powerful machines, quantum bits or qubits manipulate the superpositions to essentially perform myriad computations in parallel.

However, this free-spirited existence of quantum particles comes with its own set of challenges. Any inadvertent observation, any untamed interaction with the environment, causes what we call decoherence—the party pooper in quantum mechanics. It's the equivalent of the outside world storming into the art gallery, forcing our painting to choose between the ocean and the cityscape. When decoherence steps in, superposition is lost, and with it, the might of quantum processing is compromised.

Scientists and engineers globally are doggedly working to preserve these delicate states, endeavoring to maintain coherence long enough for quantum computations to complete their Herculean tasks. After all, a quantum computer that can't keep its qubits in superposition is like a race car with flat tires—full of potential but going nowhere fast.

The quest to conquer decoherence and harness the power of superposition is akin to learning to balance on a tightrope. It's an area ripe with exploration and excitement—a challenge beckoning the brave to redefine the bounds of technology. The scaling of quantum systems for practical applications is a race against time and interference, pushing us inexorably towards the next plateau of human ingenuity.

In confronting these challenges, quantum theory offers not just the promise of new tools and gadgets but a window into the fabric of reality itself. It prods us to question basic assumptions about the nature of particles, the architecture of space, and the tick-tock of time. The principles of superposition, chimerical as they might seem, are gradually revealing the inner workings of the cosmos.

Throughout this escapade of quantum gymnastics, what remains is the sheer elegance of a universe that plays by rules both whimsical and mathematic. As you turn the pages of this quantum adventure, you'll find the enigmas of the quantum world slowly yielding to the years of human curiosity and endeavor. Superposition isn't just a quirky quirk of the quantum world—it could be the key that unlocks the deepest secrets peppered throughout our universe.

By untangling the delicate threads of superposition, we've embarked on a journey steeped in both mystery and opportunity. It's a pilgrimage that not only reshapes our understanding of the quantum world but also redefines the landscape of technological innovation. So, while the quantum world continues to dazzle with its indeterminate state of 'this and that,' it calls on us, the seekers and sages, to persevere, to learn, and to imagine the future that these very principles will forge.

QUANTUM TELEPORTATION AND INFORMATION TRANSFER

In the quantum world, things act in ways that can seem like they're straight out of a science fiction novel. One such fascinating concept is that of

quantum entanglement. Entanglement, for many, represents the epitome of quantum weirdness. It's a phenomenon so mysterious that even Albert Einstein referred to it as "spooky action at a distance." Yet, in recent times, it's become clear that this 'spooky' property could be the bedrock of future technologies. So, what is entanglement, and why does it matter?

Imagine two particles -- let's call them Alice and Bob -- that interact in such a specific way that their quantum states are intimately linked, regardless of the distance separating them. This means the information about one instantly influences the other, no matter if they are one meter or one galaxy apart. This is quantum entanglement.

Definition and Characteristics of Entangled States

Quantum entanglement occurs when pairs or groups of particles interact in ways that the quantum state of each particle cannot be described independently of the state of the others, even when the particles are separated by a large distance. This is not just theoretical; numerous experiments have validated the phenomenon, consistently showing that the physical parameters of entangled particles, such as position, momentum, spin, and polarization, are interconnected.

The moment you measure one entangled particle, you instantly know the state of the other, a property that seems to violate the very fabric of space-time that Einstein's theory of relativity is built on. This is one of the central mysteries of quantum mechanics.

Bell's Theorem and Violation of Local Realism

The counter-intuitiveness of entanglement set the stage for physicist John Bell, who in **1964** provided a theorem that could be experimentally tested. Bell's theorem demonstrated that if the quantum mechanical predictions about entanglement are correct, then outcomes of measurements made on entangled particles are correlated in a way that cannot be explained by any theory of local hidden variables that obey the principle of locality – the idea that an object is only influenced directly by its immediate surroundings.

Put simply, Bell showed that if quantum mechanics is correct, then the measurable properties of entangled particles are determined in part by events that happened in their shared past, and not by any hidden pre-set properties or by any signal traveling between them. Experiments to test Bell's theorem have repeatedly verified the bizarre predictions of quantum physics, undermining our classical view of the world.

EPR Paradox: Einstein, Podolsky, and Rosen

In **1935**, Einstein, Podolsky, and Rosen published a paper intended to show that quantum mechanics was incomplete. They proposed a thought experiment, which came to be known as the EPR paradox, to illustrate that two entangled particles could instantly affect one another – a 'spooky action' that Einstein felt was impossible.

The EPR paradox presents two particles that have interacted such that their quantum states are entangled. If you measure one particle, you instantaneously affect the state of the second, regardless of the distance between the two. To Einstein, this implied that either information was traveling faster than the speed of light, which contradicted his theory of relativity, or that quantum mechanics was lacking some hidden variable that to date has never been found.

Einstein's discomfort with entanglement stemmed from his belief in a deterministic universe, where action at a distance without any mediating field or signal could not occur. Yet, the universe on the quantum scale doesn't seem to play by those rules.

Quantum Cryptography and Secure Communication

Quantum entanglement has profound implications for the future of secure communication. It forms the basis of quantum cryptography–a way of sending information that is immune to eavesdropping. How does it work? Quantum Key Distribution (QKD), the most famous application of quantum cryptography, uses entangled particles to generate a secret key between two parties. Since any attempt to intercept the key would disrupt the entanglement and therefore be immediately evident, QKD provides an unprecedented level of security.

In QKD, suppose Alice wants to send a secret message to Bob. They share a stream of entangled particles, with Alice keeping one from each entangled pair and sending the other to Bob. They independently measure their particles and create a key based on their results. Due to the nature of entanglement, the keys will be identical and random. If someone tries to eavesdrop, the physical state of the entangled pairs would change, alerting Alice and Bob to the intrusion.

Quantum Teleportation and Information Transfer

Entanglement also lies at the heart of quantum teleportation—a method by which the quantum state of a particle can be transmitted from one location to another, with the help of classical communication and a shared entangled pair. This doesn't mean teleporting objects in the sci-fi sense; rather, it's about transferring the exact state of a particle to another, essentially creating an identical copy at a distance.

In a laboratory, this is done by entangling two particles and sending one to a separate location. When a third particle, whose quantum state we want to teleport, is interacted with the first entangled particle, key information about its state can be sent to the second entangled particle, allowing it to take on the exact state of the third particle. This astonishing operation shows the potential mastery over the subatomic world that quantum mechanics offers.

Challenges and Future Directions in Entanglement Research

Despite the seemingly magical properties of entanglement, there are limitations. Quantum entanglement is a fragile state that requires incredibly controlled conditions to observe and maintain. Any external influences can cause decoherence, which disrupts the entangled state, making it one of the major hurdles in practical applications.

Research is ongoing in extending the distances over which entanglement can be maintained and in the development of quantum repeaters – devices that can help preserve the entanglement over long distances. Scientists are also exploring the use of entanglement in high-precision clocks and GPS systems, which could revolutionize navigation.

In conclusion, quantum entanglement, once a philosophical dilemma, is now at the forefront of cutting-edge technology and research. It pushes our understanding of the universe to new limits and paves the way for innovative applications that promise to transform the world. And while we continue to unravel its mysteries, one thing is clear– the quantum world is far weirder and wondrous than we could have ever imagined. So let us embrace the strangeness, for in it lies the potential to revolutionize the future.

CHALLENGES AND FUTURE DIRECTIONS IN ENTANGLEMENT RESEARCH

Imagine if you will, a world where we've harnessed the very whispers of nature, fine-tuning our senses to catch sounds softer than a pin drop in the bustling heart of New York city. Such is the realm we enter with quantum sensors. These futuristic devices exploit quantum phenomena, like superposition and entanglement, to measure physical quantities with jaw-dropping precision — a feat unattainable by classical means.

Delve into the unseen with me as we explore how quantum sensors and precision measurement are reshaping technology, an enthralling narrative where the micro-scale unlocks colossal potential.

Quantum Sensors: Eyes That See the Unseen

At the heart of quantum sensors lies a delicate dance between sensitivity and stability. These sophisticated instruments can detect the faintest environmental changes, including gravitational pulls, magnetic fields, and temperature, with unprecedented accuracy. It's akin to feeling the silkiness of butterfly wings without altering their flight — an intimate interaction with nature that doesn't intrude but observes, listens, and understands.

Consider the atom interferometer, a device so sensitive, it can measure gravitational variations by observing the freefall of atoms with laser-cooled precision. This sparkling ballet of particles in an optical lattice is not just a feast for the eyes but a revolution in surveying our planet's resources and monitoring volcanic activity, offering an oracle's foresight into seismic events.

Or gaze at the NV (Nitrogen-Vacancy) centers in diamonds, those tiny imperfections transformed into a lattice of spies. These fluorescent defects, a quirk of carbon's family jewels, serve as nanoscale magnetometers. Through their photon-mediated whispers, we venture into the human brain's neural odyssey, mapping magnetic fields around neurons as they fire, opening new frontiers in medical imaging and diagnostics.

Turning the Quantum Knob to Amplify Sensitivity

Precision is the cornerstone of the edifice of physics. In harnessing the quantum realm, scientists spin the knob, birthing sensors that surpass classical limitations. As we refine these devices, we amplify their sensitivity, intercepting photons without a trace, or distinguishing electric fields with the subtlety of a master perfumer separating scents.

Solid-state sensors are a testament to this, working their magic in mobile phones and navigation systems, providing a compass that points true north with quantum coherency. Such tools enhance earthquake preparedness, guide surgical instruments with microscopic fidelity, and pioneer navigation systems that lead us through the cosmos without reliance on satellite whispers.

The quantum realm is like a well of potential, from which we draw droplets of exactitude to sprinkle over the fabric of technology, weaving patterns unseen but profoundly felt. It is not just in magnification where these sensors excel, but in their resistance to noise — the cacophony of everyday life. They distinguish the symphony of the universe over the mundane chatter, where classical sensors falter, overwhelmed by the hubbub.

Precision Measurement: The Maestro's Baton

As we probe deeper into the quantum world, we find precision is everything. This is where the Maestro waves his baton, and every quantum state plays in harmony, leading to measurements defined by the fundamental constants of nature. In such a realm, even time bows down to quantum mastery.

Take, for example, atomic clocks that measure time by observing electron transitions in atoms like cesium or rubidium. These clocks don't just tick-tock; they hum to the quantum tunes of the atoms, orchestrating a temporal passage so precise that one could lose barely a second over millions of years. These marvels synchronize satellites and refine the pulse of the Internet, placing the world at our fingertips with a tap and a swipe.

Then there's the quantum Hall effect: a symphony played out on a two-dimensional electron stage, forming a standard for electrical resistance rooted in the fundamental properties of the electron and the magnetic field. This standard sing universal language and unites nations in measurement, ensuring one ampere here is an ampere there, across oceans and beyond the horizon.

Embracing Uncertainty: The Bedrock of Precision

In this symphony of the infinitesimal, embracing the uncertainty principle is not an admission of defeat but a proclamation of our proficiency. By skirting around it, we engineer devices that use the probabilistic nature of the quantum world as an asset, not a liability. It's a carnival where even randomness spells order and chaos drums to precision's beat.

Consider quantum radar: a budding technology leveraging entangled photons to reveal objects stealthier than a whisper at midnight. Traditional radar might bounce off, fooled by material trickery, but quantum radar ensnares these elusive specters, piercing through cloaks of invisibility. This isn't a spy thriller; it's the craft of our future defense systems, warding off threats unseen.

A Quantum Leap into Tomorrow

It's no overstatement to say we stand on the cusp of a revolution — a quantum leap into a future teeming with potential. As we ensnare the trickster particles, harness their dance, and read their secrets, we unlock doors previously imagined only in the wildest of science fiction pursuits.

Yet, amidst this fantastical journey, we aren't just observers: we're participants. Every advancement in quantum sensing and measurement technology is a step toward an unwritten chapter of humanity. From refining climate change models to honing drug development with atomic subtlety, we wield the quantum brush to paint our tomorrows.

Where the brushstrokes take us is a story only time will tell. But one thing is certain: with quantum sensors and precision measurement, we're drawing closer to perceiving the heartbeat of the cosmos, and that, dear reader, is a narrative where every twist, every revelation, is a marvel unto itself. Let us, therefore, march forward with eyes wide open, into the embrace of what was once cloaked in shadow, now illuminated by the quantum light.

CHAPTER 5: QUANTUM MECHANICS IN TECHNOLOGY

5.1 Quantum Mechanics in Modern Technology

Quantum Sensors and Precision Measurement

Immerse yourself in the remarkable principles of quantum computing, a field burgeoning with potential, a frontier ready to redefine every ounce of computing power we've wielded so far. What appears as an enigmatic chapter within the voluminous tome of quantum mechanics is steadily unfurling the fabric of our digital universe.

Quantum computing departs from classical computing at the most fundamental level. It's akin to comparing the flutter of a butterfly's wings to the tumultuous gale of a hurricane. Classical bits, the very alphabet of traditional computers, are binary – they exist as either zero or one. However, in the quantum realm, we encounter the quixotic quantum bit or 'qubit.'

Qubits are the heart of quantum computing. They bask in a state of superposition, a quantum limbo that allows them to be in a combination of states simultaneously. Picture a globe with poles labeled zero and one: traditional bits can be at either pole, while a qubit can exist at any point on the surface, embodying a blend of zeros and ones in a breathtaking ballet of probabilities.

When you entangle qubits, a strange and spooky kind of quantum togetherness, you set the stage for computational symphonies unseen in classical systems. Change one entangled qubit, and its partner waltzes to the same tune instantly, irrespective of the cosmic distances separating them. This 'spooky action at a distance' was a peculiarity even Einstein wrestled with. Yet, it is this unique feature that potentially enables quantum computers to solve complex problems with an almost otherworldly synchronicity.

No quantum feat could be touted without delving into interference. Here in the quantum domain, we watch as wave functions – equations describing all potential states of a system – constructively and destructively interfere with each other. This fine-tuned interference is guided algorithmically, weaving a final pattern, a solution out of the multitude of possibilities. It is the Machiavellian manipulation of interference that allows quantum algorithms to perform their magic.

Let's now turn our gaze towards a marvel within the quantum playbook – quantum algorithms. Unlike their classical counterparts, which plod ahead with a deterministic single-mindedness, quantum algorithms like Shor's algorithm for factoring large numbers, or Grover's algorithm for searching databases, are a dance of probabilities. They can explore a vast search space more efficiently, potentially reducing the time for some computations from millions of years to mere seconds.

Entitlement to these quantum wonders demands an understanding of the intricacies involved. The process of quantum error correction and quantum logic gates are challenges that must be conquered. In the quantum world, the enemy we face is decoherence – the loss of quantum behavior as our qubits interact with the environment. Quantum error correction is like a digital immune system, fighting off the errors introduced by this interaction and protecting the fragile state of our qubits.

Were it not for quantum logic gates, the functional operators of this quantum circus, the exquisite choreography of computational processes couldn't be performed. These gates manipulate qubits, changing their states, entangling them, and choreographing the steps of quantum algorithms. Operated by the gentle and precise application of energy, such as photons, these gates are what allows a quantum computer to potentially outpace their classical ancestors. But remember, this realm is nothing if not nuanced; coaxing desired behaviors from qubits through quantum gates is not a task for the faint-hearted.

The journey from theory to application is fraught with trials that are as complex as they are captivating. Quantum computing today is still in its experimental youth – a technological adolescence, if you will. Current machines, known as Noisy Intermediate-Scale Quantum (NISQ) devices, grapple with errors and the stability of quantum states.

Despite these trials, pioneers forge ahead, crafting devices with increasing numbers of qubits. IBM's Quantum Hummingbird, Google's Sycamore – these are not just curiously named technological experiments but are beacons in our odyssey toward a quantum future. The horizon promises quantum supremacy, the point where quantum computers perform tasks beyond the reach of even the most powerful classical machines.

As researchers and engineers refine these techniques, quantum scalability comes into sharper focus. We endeavor to build systems that can handle enough qubits to solve the grand challenges facing humanity. Quantum simulations could unlock mysteries from the subatomic to the cosmological scale, revolutionizing fields like drug discovery, materials science, and perhaps even unlocking new sources of energy.

Yet, startling as these advances are, quantum computing's transformative potential remains grounded by present-day hurdles. These systems demand temperatures colder than the furthest reaches of space, held in check by cryogenic engineering feats. They require shielding from cosmic interference and, above all, a symphony of scientific and technical disciplines working in concert.

What awaits us at the dawn of the quantum computing era is a renaissance of computation, with machines capable of tackling problems that we once dreamed were insurmountable. The tantalizing possibilities of medical breakthroughs, economic optimizations, and even revolutions in communication and cryptography hang in the balance, waiting for us to untangle the quantum conundrum.

In the grand scheme of technological evolution, quantum computing may still be a fledgling, but it harbors the DNA of a leviathan. The untapped potential within this quantum leap beckons us, asking us to question, learn, and envision a future where the 'impossible' becomes part of our new quantum-enhanced reality.

QUANTUM COMMUNICATION NETWORKS

In our exploration into the enthralling world of quantum mechanics, we've seen how its principles bend the very fabric of our reality. But where these ideas take a sharp turn into the realm of wonder is in the domain of quantum computing, a field poised to revolutionize the way we solve complex problems that modern supercomputers would take centuries to untangle. In this captivating juncture of innovation, let's delve into the heart of this paradigm shift: quantum algorithms and their computational advantage.

Quantum computing harnesses the peculiarities of quantum mechanics to perform calculations at speeds unattainable by classical computers. Unlike traditional bits that hold a value of either **0** or **1**, quantum bits, or qubits, luxuriate in the richness of superposition, allowing them to represent both **0** and **1** simultaneously. This might sound like a subtle shift, but it's akin to discovering a new dimension of space—it's that transformative.

Imagine boarding a plane, unsure of where you'll land. In a classical world, the plane would fly to one destination per trip—efficient, but limited. In contrast, a quantum plane could explore multiple cities at once, uncovering the best vacation spot in a fraction of the time. That is the splendor of quantum parallelism. Algorithms in this quantum realm unlock the full potential of this ability, finding routes through computational landscapes that classical algorithms can't even map, let alone traverse.

One shining example of such an algorithm is Shor's algorithm, introduced in **1994**, which factors large integers exponentially faster than the best-known classical algorithms. This isn't just a neat party trick; it's a piercing arrow aimed at the heart of current encryption methods. RSA, a cornerstone of internet security, relies on the laborious task of factoring large primes to keep data safe. Shor's algorithm, in a quantum world, reduces this Herculean task to child's play, demanding a reconceptualization of digital security.

Another marvel in the quantum algorithmic tapestry is Grover's algorithm. It accelerates the search through unsorted databases—a task as common as searching for a book in a library without a catalog. Classically, this is a game of patience, examining each item until you stumble upon the one you seek. Grover's algorithm, however, shortens the search process quadratically, proving once again the nimbleness of quantum computation.

The computational advantage bestowed by these algorithms is not confined to a few esoteric tasks; it touches upon a multitude of areas, like simulating molecular behavior for drug discovery or optimizing complex systems such as supply chains or traffic flows. Here, quantum computers act like a Swiss Army knife for intricate problems, each algorithm a specialized tool folded within its innards, ready to be deployed for tasks once deemed unmanageable.

At this juncture, some may wonder how these quantum marvels come to life. Enter quantum circuitry, a symphony of logical gates, different from their classical counterparts, designed to exploit coherence and entanglement—those mysterious quantum threads that weave our universe together. Through a meticulously orchestrated sequence of operations, quantum circuits manipulate qubits, transforming quantum states and elegantly executing algorithms.

The impact of this computational might reaches beyond mere speed. It's the difference between trudging through a snowy path on foot and gliding over it in a sleigh, the cold biting at your face as you're empowered to conquer vast distances that would otherwise be insurmountable. It's not just doing things faster; it's achieving the previously unachievable.

Yet, as with any frontier, challenges abound. Implementing these algorithms requires navigating the churning waters of decoherence and noise, ensuring that our quantum states don't lose their mystical connections to the environment before their job is done. Current quantum hardware is like the early flying machines—a glimpse into a soaring future, but with a journey still susceptible to the occasional tumble. Taming this wild quantum landscape is the task of a pioneering generation of quantum engineers and theorists.

In the practical world, these hurdles mean that there's a race towards what's termed "quantum supremacy"—the point where a quantum computer can perform a task no classical computer can. In **2019**, for the first time, this breakthrough was claimed by achieving a calculation in a realm where classical computers falter. It was a proof of concept, not yet a tool at our disposal but a beacon illuminating the way forward.

As this new era dawns upon us, businesses and governments are already positioning themselves on the starting lines, ready to capitalize on the quantum leap in processing power. The race isn't simply for bragging rights; it's a strategic maneuver in a world where computational might equates to economic, scientific, and national potency.

To circle back to you, the astute reader and future architect of the quantum era, the array of algorithms we've skimmed paints a portrait of a world brimming with potential, waiting for the curious and the bold to unravel its mysteries. These algorithms are the keys to unlocking the computational advantage that will reshape industries and redefine our capabilities, empowering us to tackle the grand challenges of our world.

As we wind down this whirlwind tour of quantum algorithms and computational advantage, it's pivotal to recall that quantum computing isn't simply about outpacing what's come before. It's a transformative tool, expanding our understanding, enabling new science, and accelerating innovation. We're standing at the precipice of a computational revolution, and as you absorb these insights, you're not just an observer; you're a pioneer in this quantum voyage, exploring the contours of a future shaped by the principles of a world that's both shockingly strange and endlessly fascinating.

QUANTUM SENSING AND IMAGING TECHNIQUES

Quantum computing stands as a testament to human ingenuity, a transformative force poised to redefine the landscape of technology and problem-solving. Its inception springs from the heart of quantum mechanics, a domain where particles exist in multiple states simultaneously and where the traditional limits of computation are replaced by a symphony of probabilities.

The premise of quantum computing revolves around the quantum bit, or qubit—a unit of quantum information that, unlike classical bits, isn't restricted to being just a **0** or a **1**. Instead, it can exist in a superposition, embodying both states at once until measured. Imagine the vast parallel universes of computation that unfurl from this single peculiarity. A quantum computer with just **50** qubits could process more data than there are atoms in the known universe. This is not hyperbole; it's the bizarre and mesmerizing promise of quantum technology.

As with any revolutionary technology, quantum computing is rife with both challenges and opportunities. As disruptive as it is, it invites us to soar, but reminds us to cross storms to reach new horizons.

Deciphering Quantum Complexity

The labyrinth of quantum computing is complex, not just in the theoretical underpinnings of quantum mechanics but in the physical realization of a quantum system. Qubits are capricious entities, prone to losing coherence through interactions with their environment—a pitfall known as decoherence. In a mere fraction of a second, the rich quantum information can dissipate, turning our sophisticated quantum machine into no more than a heap of classical ruins.

To combat this, researchers are devising ingenious methods of error correction and fault tolerance. These techniques are akin to the insulated walls of a pin-drop silent room, guarding the delicate whispers of quantum information from the cacophony outside. Yet, the insulating materials we have at hand are still imperfect, and the quest for stable qubits at scale continues to galvanize universities and tech giants alike. With each stride towards better qubit stability, the door to reliable quantum computing swings wider open.

Quantum Supremacy and Beyond

The term "quantum supremacy" heralds an era where quantum computers solve problems that classical computers, for all their speed, could never unravel in a human lifetime. Picture solving intricate molecular structures for drug discovery, breaking cryptographic codes that anchor digital security, or simulating quantum physics itself—tasks that would leave conventional supercomputers beleaguered.

However, quantum supremacy isn't the endgame; it's merely the beginning. As quantum computers teeter on the brink of this milestone, we delve deeper into the algorithms designed for a post-supremacy world. These algorithms serve as the conduit between quantum potential and practical application, guiding us from abstract qubit gymnastics to tangible solutions that ripple through every tier of science and industry.

Unleashing Computational Power

In domains like optimization and machine learning, the computational prowess of quantum computing could be nothing short of revolutionary. Consider logistic companies optimizing complex delivery routes or financial analysts crunching the probabilistic permutations of markets. Quantum computers, operating at their hypothetical zenith, could drastically compress timelines for these gargantuan tasks, transforming decision-making from ponderous to almost precognitive.

Yet, the code that would marshal this computational power remains in embryonic form. Quantum programmers of tomorrow must be fluent in the language of qubits, entanglement, and superposition. Envisioning a thriving ecosystem of quantum software requires investment in education and tooling to arm a new wave of developers with the quantum lexicon.

Scaling the Quantum Summit

Scaling a quantum computer from a proof-of-concept to a commercial colossus involves more than simply amassing qubits. The architecture of quantum systems must be holistically designed to harmonize with the delicate dance of quantum states. Quantum processors have to operate in unison, achieving an interconnected web of computational choreography. This feat demands innovations in fabrication, cryogenics, and quantum error correction—each a specialized field in its own right.

Integrating these disparate elements into a functional quantum computer is akin to building the first flight-worthy airplane. The Wright brothers didn't just strap wings to a bicycle; they wrought a harmony between propulsion, aerodynamics, and control. Quantum engineers today face a similar epoch-defining challenge.

Ethical Frameworks and Societal Impact

As with any disruptive technology, quantum computing wades into uncharted ethical waters. The very strength of quantum computation—the ability to crack contemporary encryption casts a shadow of vulnerability over data security. Industries and governments must preemptively address these risks, weaving an ethical fabric robust enough to cushion the impacts of quantum disruption on privacy, security, and economic stability.

Moreover, quantum computing stands to exacerbate societal inequalities if left unchecked. Access to these rarefied layers of computational power cannot become the preserve of a privileged few. Ensuring that the fruits of quantum research benefit society at large involves concerted efforts in policy, education, and open-source initiatives.

Quantum Leap or Quantum Mirage?

The journey towards quantum computing is fraught with both unbridled potential and sobering pitfalls. Encountering quantum computing where it stands today—perched on the precipice of practicality—is akin to witnessing the assembly of the first electrical grid or the ignition of the first silicon chipset. We are custodians of a quantum seed that, if tended with care, could grow into a tree whose branches spread into every digital crevice of society.

James Quantum beckons you to view this not as distant science fiction but as a living narrative where you play a pivotal role. Together, we stand on the cusp of a quantum paradigm shift—a shift that rewires our digital future and redefines what's possible. Are we ready for the quantum leap? The promise is vast, and the tapestry of quantum computing continues to weave—an intricate interplay of challenges and boundless opportunities that only the bold will embrace and explore.

5.2 Quantum Computing: A Paradigm Shift

Principles of Quantum Computing

In an ever-expanding universe, small and big, one of the most enthralling mysteries that continue to captivate scientists and laypeople alike is the notion of quantum entanglement. This remarkable quantum phenomenon defies our classical understanding of the world and raises profound questions about the very nature of reality.

Quantum Entanglement and Spooky Action at a Distance

Imagine two particles that are entangled – a condition so mystically intertwined that the state of one instantly influences the state of the other, no matter the distance between them. This peculiar connection was one Einstein famously referred to as "spooky action at a distance," a phenomenon he himself was hesitant to accept as it appeared to challenge the universal speed limit—the speed of light.

Entanglement commences when two or more particles become correlated in such a manner that the measurement of one particle's quantum state—such as its position, momentum, spin, or polarization—directly influences the state of the other particle, this instantaneous connection seemingly transcending the barriers of space and time.

The perplexing nature of quantum entanglement is grounded in the principle that quantum particles can exist in a superposition of states. When a measurement occurs, and the particle's state is 'chosen,' any other particle entangled with it seems to 'know' about this decision instantly and assumes a corresponding state that is necessarily defined by its counterpart's state, despite any separation in distance. This is not just theoretical musing—quantum entanglement has been experimentally proven, driving home the point that particles can become entangled, challenging the very notions of locality and causality.

Scientists conduct tests on quantum entanglement utilizing a concept called Bell's Theorem. This theorem provides a way to test whether particles are entangled by examining whether the predictions of quantum mechanics concerning their behavior are more accurate than those made by classical physics, based on local hidden variables. Time and again, the theorem leans in favor of quantum mechanics—validating entanglement and shaking the foundations of classical physics.

But it was the EPR Paradox, proposed by Einstein, Podolsky, and Rosen, that brought the strangeness of quantum mechanics into sharp focus. They posited that either the information was traveling faster than light or, alternatively, the characteristics of one particle were pre-determined, remaining hidden until measured. While Einstein yearned for a deterministic explanation, the seemingly teleportational attribute of entanglement hints at a universe less beholden to our everyday experiences and intuitions than previously thought.

The implications of entanglement burgeon well beyond theoretical curiosity. It is at the center of numerous cutting-edge technologies. Quantum cryptography, for instance, employs the principles of entanglement to create unbreakable encryption. In this cryptographic approach, any attempt to eavesdrop on a communication would disturb the entangled particles and be immediately noticeable due to the inseparable correlation between them.

Even more mind-bending is the concept of quantum teleportation, which exploits entanglement to transmit quantum information—such as the exact state of a particle—from one location to another, without the need for physical transport of the particle itself. While still in its nascent stages, quantum teleportation could revolutionize the way information is sent, potentially leading to the development of a quantum internet that would elevate the security and speed of data transmission to previously unimaginable heights.

Moreover, quantum entanglement is not just a curious feature to be harnessed for future technologies. It is an essential component in the most ambitious quantum computers. These marvels of technology utilize the entanglement of quantum bits, or qubits, to perform computations at speeds that would leave classical computers far behind.

However, precisely controlling entangled particles across macroscopic distances presents a significant challenge for researchers. Maintaining entanglement involves shielding particles from any interference, lest the entanglement breaks—a phenomenon known as decoherence. This delicate process remains one of the substantial hurdles to harnessing the full potential of quantum entanglement in everyday applications.

While the applications of quantum entanglement promise to be transformative, there are still formidable technical and theoretical difficulties that remain to be surmounted. These include not just practical limitations such as maintaining the delicate state of entangled particles over long distances, but also deep philosophical and scientific questions about the nature of reality itself.

The future research into quantum entanglement will no doubt continue to shake the very foundations on which our classical views of the universe are built. As technologies mature and our understanding deepens, the strange quantum threads that tie the universe together might just turn out to be lifelines into a new era of technological advancement and theoretical insight.

In your journey through the quantum world, as you ponder the remarkable union between entangled particles, let yourself be inspired by this emblematic testament to the quantum realm's enigmatic beauty—a world where distance is but an illusion and where the instantaneous interplay of the cosmos comes to light in the dance of entanglement.

QUANTUM ALGORITHMS AND COMPUTATIONAL ADVANTAGE

Quantum mechanics is enigmatic—a plunge into a world far removed from the classical intuition accumulated over our lifetimes. For many, myself included, this journey began with a sense of befuddlement, a dash of disbelief, and a generous measure of awe. But fear not, for the quantum realm, as complex as it may seem, is a trove of possibilities that root our most sophisticated innovations and technologies—yes, even the screen you're reading this on might owe its existence to quantum principles.

Imagine a future where quantum sensors detect diseases at their inception or quantum networks provide unhackable communication channels. This isn't a snippet from a science fiction novel; these are the precipices of our current technological evolution. Decoding quantum mechanics offers us a key to unlock these advancements, an adventure I invite you to join as we explore how quantum mechanics operates at the very edge of our everyday life and beyond.

To envision the impact of quantum mechanics, consider its strangest feature affecting our view of reality: superposition. At first, it appears as if quantum objects can be in multiple states at once—a notion that contradicts everyday experiences. Nonetheless, this bizarre behavior is the cornerstone of quantum computing, a rapidly developing field anticipated to revolutionize problem-solving capacities in industries, security, and scientific research.

Quantum computers leverage the power of superposition to perform calculations exponentially faster than their classical counterparts. Let's delve into this by imagining a traditional computer, operating with a binary system where the fundamental units of data—bits—are either **0**s or **1**s. In stark contrast, quantum computers utilize qubits, which can exist as **0**, **1**, or any superposition of these states. It's like flipping a coin and having it whirl in a suspended state of heads and tails simultaneously, enabling the computation of multiple possibilities at once.

The potential here is not merely incremental but exponential. Tasks like factoring large numbers, which would take classical computers longer than the age of the universe, could be accomplished in a fraction of the time. Quantum algorithms, like Shor's algorithm for prime factorization, exemplify this potential and could one day render current encryption methods obsolete, prompting a complete overhaul of cybersecurity strategies.

However, a world equipped with such powerful tools also bears significant responsibility. Quantum encryption, for instance, offers the promise of secure communication channels impervious to traditional hacking attempts. Quantum key distribution (QKD) utilizes the principles of quantum mechanics to ensure that the key to decrypt a message cannot be intercepted without detection. This "quantum leap" in security is poised to change how we think about privacy and information sharing on a global scale.

Yet, with great power comes great challenges. Decoherence, the enemy of quantum coherence, arises from the interaction of a quantum system with its environment, resulting in the loss of superpositional states. Scientists and engineers are on a constant quest to isolate quantum systems more effectively to preserve this fragile state necessary for quantum computation and communication.

Another hurdle is scalability. Building a quantum computer with enough qubits to solve complex, real-world problems is akin to scaling a sheer cliff wearing slippers. For quantum computers to be practical, they require thousands, if not millions, of qubits operating in unison—a feat we are yet to achieve.

But let's remember that every momentous journey in humanity's history began with a single step filled with uncertainty and the shimmer of possibility. The journey toward harnessing the full potential of quantum computing and its allied technologies is no different.

Beyond computing, quantum mechanics is transforming the very fabric of technology. Quantum sensors, with their unprecedented sensitivity, are enabling new frontiers in measurement and imaging. These devices wield the superposition principle to detect minute changes in various physical quantities like gravitational fields—a boon for applications ranging from navigation to geological surveying.

Quantum metrology stretches the limits of accuracy in measurement, impacting everything from the synchronization of GPS systems to the precision of timekeeping. This level of detail could enable us to navigate the cosmos with unheard of precision, or to diagnose medical conditions long before they manifest into symptoms.

Speaking of medical applications, the future could see the deployment of quantum sensors capable of imaging the body at a cellular or even molecular level. Such quantum-enhanced technologies have the potential to revolutionize diagnostics and treatment, opening doors to early detection of illnesses that currently slip by unnoticed until they escalate.

In the realm of materials science, quantum mechanics is ushering in an era of 'quantum materials'—substances with properties emerging from quantum mechanical features like entanglement and superposition. These materials exhibit behavior entirely different from conventional materials, with promising applications in electronics, energy technologies, and beyond.

As researchers and developers, we face the daunting task of integrating these phenomena into everyday technology. But with every challenge surmounted, the promise of a quantum revolution in engineering, computing, and the physical sciences grows ever closer.

As we stand at the crossroads of a quantum-technological transformation, we also face ethical and societal considerations. The pursuit of progress must go hand in hand with a commitment to responsible innovation, ensuring that the powerful tools borne from quantum mechanics are wielded to benefit human society at large, fostering an equitable and secure future.

And so, my fellow explorers, the impact of quantum mechanics on technology is not merely a possibility—it is the unfolding story of our time. As we venture deeper into understanding and harnessing these quantum phenomena, we stand to witness and contribute to one of the most exciting chapters in the history of science and technology. The quantum age beckons, and it is our curiosity, our ingenuity, and our spirit of collaboration that will illuminate the path forward.

CHALLENGES AND OPPORTUNITIES IN QUANTUM COMPUTING

Peering into the heart of existence, where time itself coils up to the merest speck of a beginning, quantum cosmology stands as our flashlight into this profound darkness. It's where the fascinating principles of quantum mechanics court the grand structures of the cosmos, weaving the most intricate tale—the origin of the universe.

Imagine, if you will, the universe as a grand symphony, with its current expansiveness and energy merely the crescendo following a delicate and subtle introduction. Quantum cosmology studies this introduction—the earliest moments of the universe, where classical theories like the Big Bang scrape their knuckles on the ceiling of understanding. Here, the laws of quantum mechanics dance elegantly and chaotically to choreograph the very inception of space and time.

In the prevailing Big Bang model, there's a hypothetical singularity, a point of infinite density and temperature. But just as general relativity and quantum mechanics often view the world through different lenses, the singularity proves a sore spot—a place where the laws of physics as we understand them collapse under their own gravity. Notably, quantum mechanics never favored infinite anything. It suggests that there's a finite limit to how closely we can pack energy and matter.

Enter the Planck era, named after Max Planck, the German physicist who founded quantum theory. This was the universe at its first tick, **10^{-43}** seconds after the purported Big Bang, where temperatures and energies were so high that the distinction between the forces of nature blurred into obscurity. Quantum gravity, the elusive theory that seeks to reconcile gravity with quantum mechanics, states that at the Planck scale, space and time may not operate in the smooth continuum that Einstein's relativity implies but instead may be granular, made of 'quanta' of space-time.

We are introduced to the concept of quantum fluctuations. In your average day, quantum fluctuations are of little consequence—but not so in the dense, high-energy environment of the early universe. There, these fluctuations could have magnified, inflating like balloons to become the seeds of galaxies and the vast, web-like structure of the cosmos we observe today. This offers a potential solution to the horizon problem, explaining how regions of the universe that should not have had time to interact are yet so homogeneously distributed.

These fluctuations gave the universe its texture, producing tiny variations in temperature and density that we've detected in the Cosmic Microwave Background (CMB)—the afterglow of the universe's hot, dense youth, stretched thin and cooled by the relentless expansion of the cosmos. Measurements of the CMB not only support the Big Bang model but also underscore the importance of quantum processes in shaping the early universe.

At the center of this quantum beginnings story is cosmic inflation—a rapid expansion of space-time that ironed out the geometry of the cosmos. It took the dense point of the Big Bang, with its quantum idiosyncrasies, and stretched it to epic proportions. If this sounds like a surge worthy of a superhero movie, it's because the concept is almost as mind-bending as quantum mechanics itself. Inflationary theory argues this burst happened faster than the speed of light (space itself can do this without breaking physical laws) and ended as abruptly as it began.

Another triumph of quantum cosmology is the prediction of a flat universe—that is, globally, space is not curved significantly. Observations of the CMB fit snugly with this foresight, depicting a universe that balances precisely on the knife-edge of density needed to avoid collapsing back into a hot, dense state or expanding into chilly oblivion.

Yet, quantum cosmology doesn't merely trace back the origins; it offers a kaleidoscope through which to view the fate and features of the universe. It ponders on mysteries such as the eventual heat death, the possibility of multiverses, and whether our universe is one of many bubbles frothing in a cosmic pot.

This microscopic-macroscopic dialogue points to quantum gravity—a still-incomplete theory with multiple candidates, string theory and loop quantum gravity among them. While string theory suggests our universe is built upon subatomic strings vibrating in **10** or more dimensions, loop quantum gravity proposes that space-time itself is woven from loops and networks at the quantum level.

But amid all this complexity, why should these quantum origins concern us, dwellers of the macro world? Because they are a testament to the enduring canvas upon which our cosmic home is painted. Every star, every planet, every grain of cosmic dust shares this quantum birthright. It reveals the universe as a synergistic masterpiece where the vast and the void are as interconnected as particles entangled across light-years.

What marvels this quest for quantum cosmology has served up: an understanding that our universe, vast and intricate, was once cradled in the bosom of quantum processes, that the bulky bodies of galaxies were once whispers of quantum noise. In hindsight, we have peered beyond the curtain of the cosmos, beyond Einstein's universe, to a world that is continually vibrant, puzzling, and profoundly subtle.

Though this narrative skirts on the edges of philosophical musings, it is driven by rigorous mathematical models and empirical observations.

Quantum cosmology stands not simply as a field of scientific study but as a testament to human curiosity—the insatiable desire to know where we come from and, indeed, where we are headed. It is the nesting ground of our most profound existential queries and their eventual answers.

In the words of Niels Bohr, a pioneer of quantum mechanics, "Everything we call real is made of things that cannot be regarded as real." As we harness the paradoxes and lensing of quantum mechanics to view our cosmic nursery, we might realize that our universe is not just stranger than we imagine—it may be stranger than we can imagine. Quantum cosmology, therefore, is more than a quest for knowledge; it's an invitation to marvel at the quantum origins of the universe—an endless well of wonder for the curious, the seekers, and the star-gazers.

CHAPTER 6: QUANTUM PHYSICS AND THE COSMOS

6.1 QUANTUM COSMOLOGY: EXPLORING THE UNIVERSE AT THE QUANTUM SCALE

QUANTUM ORIGINS OF THE UNIVERSE

Embarking on the infinitely enigmatic journey within the cosmos, we find ourselves face-to-face with one of the universe's most secretive confidants, the black hole. An object of such dense constitution that not even light, the speediest traveler of the cosmos, can escape its gravitational clutches. But where does quantum mechanics, with its propensity for the very small, intersect with these colossal behemoths of space? Let's delve further and discover the intriguing connection.

Picture the event horizon of a black hole – it's not a surface you can touch, but rather a point of no return. It's here in this realm of extremities that Stephen Hawking, a visionary theoretical physicist, contemplated the fate of quantum particles. Adhering to the classic tenets of quantum theory, he proposed that pairs of virtual particles could spontaneously appear, as they are wont to do in the vacuum of space, near this ominous border.

In a rather extraordinary twist, if one particle gets captured by the black hole's gravity while its partner escapes, the escapee becomes real – contributing to a glow we now call Hawking radiation. This fanciful narrative might seem the stuff of science fiction, but it's a compelling prediction of quantum field theory applied to the curved spacetime around a black hole.

Now, let's unwind the spool of implications here. Hawking radiation is a remarkable bridge between general relativity (the science of the very large) and quantum mechanics (the realm of the very small). Hawking's work illuminated a pathway that could lead us to the holy grail of physics: a grand unified theory that reconciles the two.

One might wonder, with the incredulity of a sane mind, how these minuscule particles can have any measurable impact on the immensity that is a black hole? Here's where the conservation of energy prances onto the stage. As these particles siphon away the energy of the black hole, they do something extraordinary – they cause the black hole to shrink. Indeed, Hawking radiation suggests that black holes aren't eternal; they evaporate, losing mass over what is, admittedly, an unfathomable expanse of time.

The concept of black hole evaporation tantalizes us with a quantum conundrum: the information paradox. According to quantum mechanics, information can't be destroyed. But if a black hole eventually evaporates entirely, what becomes of the information about the objects it swallowed? This question stirs the pot of theoretical discourse, with proposed solutions ranging from information being stored at the event horizon to it being emitted with Hawking radiation, or perhaps it finds a new cosmic lease of life in an alternate universe.

Shift gears now and consider what these quantum gymnastics mean for our comprehension of the larger cosmos. The seeds of black holes are believed to sprout in the aftermath of supernova explosions – a stellar finale of epic proportions. These seeds, if they are sufficiently fecund, grow by feasting on the cosmic buffet of dust and gas or amalgamating with other black holes. Amid the deep dark, they could be responsible for steering the cosmic evolutionary tale, influencing the structure of galaxies with their formidable gravitational pull.

But there's more – imagine cosmic dawn, when the universe was just a hatchling. Herewithin, minuscule black holes may have formed aplenty –

theoretically known as primordial black holes. If Hawking radiation exists, these pint-sized cosmic artifacts could be fizzling out just about now, liberating energy that might just be detectable if our instruments are acute enough. They could very well contribute to the dark matter that threads through our cosmos, an enigma that binds galaxies yet lies beyond direct observation.

Now, what about our quantum explorations helps us penetrate these cosmic mysteries? Cutting-edge telescopes scanning the high-energy vista of our universe could capture the dying gasp of evaporating primordial black holes or the lighthouses that are supermassive black holes devouring stellar meals. But we don't just observe; we simulate. Quantum computing proposes to muscle through the computational juggernauts of modeling these violent, high-gravity phenomena, with nary a sweat. It holds the potential to decode the signals engraved in gravitational waves cast forth by colliding black holes – ripples in the fabric of spacetime itself that reach out to our LIGO and Virgo detectors.

Speaking of ripples, the inflationary universe theory postulates that quantum fluctuations during the universe's infantile, exponential growth spurt could be the seedlings of the sprawling cosmic web observed today. What began as subatomic jitteriness might have ballooned into the vast voids and gargantuan galaxy clusters we see. Without quantum mechanics, this monumental architecture of the universe would remain unintelligible – yet another testament to how quantum effects inscribe their signature across the cosmos.

In the grand cosmic scale, quantum mechanics emerges not just as a theory for the incredibly tiny, but as a fundamental actor in the theater of the universe. From the flickering Hawking radiation from black holes to the embryonic murmurs of quantum fluctuations seeding cosmic structure, our universe finds quantum interwoven in its very fabric.

The future beams bright with the promise of quantum technology lending us the tools to probe deeper into the nature of black holes and the quantum realm. Exciting times lie ahead, so buckle up! As we continue to extend our quantum scaffolding further into the cosmos, the thinly veiled secrets of our universe become ever so slightly more within our grasp. The story of quantum mechanics and astrophysics is being written by tireless researchers and, perhaps, by you, dear reader, as we venture into this beautifully enigmatic quantum cosmos together.

QUANTUM GRAVITY AND THE SEARCH FOR A UNIFIED THEORY

In the tranquil twilight of the cosmos, where silence reigns supreme, there's a restless dance taking place at a scale so minute, it defies the very intuition we've nurtured through our everyday experiences. Here, within the fabric of space-time, *quantum fluctuations* are the spontaneous, frenetic tremors orchestrated by the laws of quantum mechanics, churning the cosmic pot from which the marvels of the universe emerge.

It might seem fanciful, this notion that the vast expanse of space we gaze upon on a starlit night could be intimately connected to the erratic quirks of quantum phenomena. Yet, in the bizarre realm of the very small, where quantum effects dominate, these fluctuations are not just theoretical musings; they are the whispers of creation itself, laying out an intricate story of how the universe bloomed from an infinitesimal seed during the period known as *cosmic inflation*.

Imagine, if you will, a sea of potentiality, where virtual particles wink in and out of existence like fireflies in the dark. They're ephemeral, yes, but their fleeting lives have profound implications. In the vacuum of space, there is no such thing as nothing. This vacuum is alive with these quantum fluctuations, a ceaseless hum of possible particles and antiparticles generating and annihilating, giving and taking energy, even if just for a moment. It's a strange thought, isn't it, that the vacuum isn't empty but instead bubbling with these invisible possibilities?

As you dive into these quantum waves, you realize they are not confined to the realm of the infinitely small. Indeed, under certain conditions, they can stretch to cosmic proportions. This is where we make the quantum leap from micro to macro, for it was during an inconceivably short and unimaginably energetic burst of expansion—cosmic inflation—that these quantum fluctuations were magnified to the scale of the universe. Inflation proposes that in the mere **10^{-32}** seconds following the Big Bang, the universe expanded exponentially from a space smaller than a subatomic particle to the size of a grapefruit.

Why does this matter? Because the quantum fluctuations we discussed, those ripples in the void, were also stretched during this period of inflation. They transformed into the seeds of the large-scale structures we observe today—galaxies, clusters of galaxies, and vast cosmic webs. These ripples became the blueprints for the gravitational scaffolding upon which matter accumulated, the initial unevenness in density that would become, billions of years later, the cosmic landmarks we recognize.

Cosmic inflation is a bridge between the quantum and the cosmic, a tale of unity spun by the threads of theory and observation. It's astounding to ponder that every star, planet, and galaxy traces its lineage back to these quantum perturbations. Here, at the dawn of everything, quantum mechanics does not just predict the existence of particles; it foretells the structure of the cosmos itself.

Yet it is not without its enigmas. For instance, what caused inflation to start and then stop so abruptly? What mysterious field—the so-called *inflaton field*—was responsible for this tremendous expansion? These questions remain at the cutting edge of cosmological research, the answers to which may reshape our understanding of the universe's beginnings and its ultimate fate.

Cosmic inflation also tantalizingly hints at the existence of gravitational waves—ripples in the fabric of spacetime. These waves, born from quantum fluctuations amplified by inflation, could potentially carry the echoes from the birth of time itself. They traverse the cosmos, perhaps to someday be detected by our instruments, acting as fossils of the high-energy events occurring at the onset of inflation.

Moreover, the concept of quantum fluctuations during inflation nudges us towards a revolutionary idea: that our universe may just be one of many. If these fluctuations can produce a cosmos as vast and complex as our own, could it be that there are countless other universes birthed from the same quantum foam, each with its own physics, stars, and perhaps even life? This is the multiverse hypothesis—a dizzying thought that out of the quantum chaos could emerge a landscape of parallel realities.

As we ponder these ideas, the connection between the esoteric and the expansive becomes clear. Far from being confined to the realm of the theoretical, quantum mechanics has, quite literally, had a hand in sculpting everything we hold dear in this celestial tapestry. It invites us to consider our place in the universe not merely as passive observers but as progeny—and participants—in a cosmic quantum dance that spans the scales from the subatomic to the astral.

Grasping these quantum roots of cosmic phenomena not only satiates our human thirst for knowledge but also inspires a sense of awe at the intricate interplay of the laws of nature. It beckons us to continue our exploration, both at the quantum and cosmic frontiers, for it is at this intersection that we may find the keys to understanding the most fundamental aspects of our existence.

And as we march forward, straddling the line between observable reality and theoretical possibility, we must acknowledge the dual role of quantum mechanics—as both the sculptor of the universe's grand design and as the tool we use to uncover its deepest truths. It is here, within this cosmic crucible, that quantum fluctuations and cosmic inflation beckon us to peer ever closer into the quantum realm, daring us to decipher the whispers of creation that echo throughout the cosmos.

QUANTUM COSMOLOGICAL MODELS AND OBSERVATIONAL TESTS

Peering out into the night sky, we're not merely looking at stars; we're witnessing a cosmic dance choreographed by the laws of quantum mechanics playing out on an intergalactic stage. While the vastness of the cosmos seems to dwarf us, and our problems, it's the strange and tiny quantum effects that weave the very fabric of our universe. When we delve into the realm of astrophysics and cosmology through the quantum lens, we pull back the celestial curtains to reveal how quantum mechanics influences, and possibly underpins, the cosmos.

At the quantum level, our universe seems almost magical. This is a place where particles pop in and out of existence, where the behavior of one particle can instantly affect another light-years away, and where the certainty of reality itself is a blur. Understanding these effects is not just a matter of satisfying our curiosity—it's about framing the context in which every other cosmic event takes place.

Let's start with something that has captivated humanity since we first looked upward: black holes. Once a mere theoretical curiosity, black holes are now believed to punctuate the fabric of spacetime throughout the universe. Quantum mechanics brings a new layer of understanding to these enigmatic phenomena, primarily through Hawking radiation. In the **1970s**, Stephen Hawking proposed that black holes could emit radiation due to quantum effects near the event horizon, the point at which nothing, not even light, can escape the gravitational pull of a black hole. This concept suggested that black holes aren't eternal and can slowly evaporate over unimaginable time scales. Hawking's theoretical breakthrough melded the realms of relativity and quantum theory and hinted at a deep connection that we're only beginning to unravel. Understanding this emission mechanism helps us grapple with the ultimate fate of black holes and, possibly, the universe itself.

On a larger scale, quantum fluctuations are the minuscule temporary changes in energy that occur in empty space. This goes against everything our intuition tells us—that 'nothing' can't have properties or do anything at all. However, in the quantum vacuum, 'nothing' is, in fact, a seething cauldron of activity. These fluctuations may have played a monumental role during the earliest moments of our universe's existence, during the period of cosmic inflation. This was when the universe expanded at an incomprehensible rate, faster than the speed of light, immediately after the Big Bang. It's these quantum fluctuations that could have provided the seeds for the large-scale structure of the universe we see today, the galaxies, the clusters, all tracing back to tiny quantum blips in the very early cosmos.

One of the most fascinating implications of quantum effects in cosmology is the search for a unified theory. Quantum mechanics and general relativity, Einstein's theory of gravity, are the two pillars upon which our understanding of physics rests. Yet, they are fundamentally incompatible at their core. While general relativity beautifully describes the macroscopic world, including the gravity-bending dance of celestial bodies, quantum mechanics governs the unpredictable and often nonsensical behavior of the microscopic world. A unified theory—often posited in the form of Quantum Gravity—would be the Holy Grail of physics, harmonizing these two realms into a single framework. The quest for understanding these fundamental forces as expressions of the same underlying principles could lead to profound revelations about the origins and ultimate fate of the universe.

Exciting times lie ahead as we seek observables that tie quantum mechanics to cosmological phenomena. The recent observation of gravitational waves, ripples in spacetime predicted by general relativity, not only confirmed a century-old prediction but also opened up a new avenue for exploring the universe. The nascent field of gravitational wave astronomy may one day give us a window into the quantum aspects of gravity and the early universe.

Additionally, the study of cosmic microwave background radiation, the echo of the Big Bang, has provided us with a cosmic map of the early universe's quantum fluctuations. With each subtle temperature variation, we read the historical quantum tremors that gave structure to everything we see in the night sky today.

Perhaps most compellingly, the quantum cosmological models we're developing are ripe for observational tests. These models predict specific signatures that could be observed in the cosmic microwave background or in the distribution of galaxies across the cosmos. Although the scales involved are mind-boggling, with sufficient technological innovation, we might just translate those intricate quantum variations into observable evidence.

As we contemplate our place in the universe, we must acknowledge that the very act of observation, so critical in quantum mechanics, is not limited to laboratories. Every time we gaze out into space, every time we capture a photon that has traveled unimaginable distances, we are actively engaging with the quantum underpinnings of the cosmos.

Understanding the cosmos through the quantum lens does not render the universe cold or mechanical; rather, it imbues it with a layer of mystique and a promise of deeper connections waiting to be discovered. It reassures us that no matter the scale—be it the immensely large or the infinitesimally small—the principles that govern us are intricately linked. Quantum mechanics, with its non-intuitive quirks, does not isolate us, but rather connects us intimately to the vast narrative tracing back to the birth of time itself.

In the end, quantum mechanics and the cosmos perform an elegant celestial ballet, reminding us that we are part of a universe far more fantastic than anything we could have imagined. Our journey through the quantum cosmos is just beginning, and the revelations to come will undoubtedly transform our understanding of reality and our place within it. The cosmos is not simply something to be observed—it is something to be understood, and quantum mechanics is the key that unlocks that understanding.

6.2 Quantum Effects in Astrophysics and Cosmology

BLACK HOLES AND HAWKING RADIATION

Imagine awakening in a world where every object has a peculiar ability to be in more than one place at the same time, or where a cat in a box could be simultaneously alive and dead until someone peeks inside. This isn't the stuff of fairy tales—it's the subatomic fabric of reality, woven by the threads of quantum superposition and measurement.

Quantum superposition, a concept that can twist the mind more adeptly than any pretzel, posits that particles like electrons and photons exist in multiple states or locations at the same time—until they are observed. To untangle this mystery, we turn to the iconic Schrödinger's cat thought experiment. Envision a feline sealed in a box alongside a radioactive atom and poison that's released when the atom decays. Quantum mechanics suggests the cat is both alive and dead in a superposed state until the box is opened. While this sounds implausible, it simply indicates how the principles governing atomic and subatomic particles seem out of place in our macroscopic daily existence.

This feline paradox illustrates a profound question: how do quantum states evolve? According to the pioneering Schrödinger's equation, these states change smoothly over time. However, this serene evolution halts abruptly during a measurement—the wave function, which encodes the probabilities of finding a particle in any particular state, collapses to a definitive outcome. This collapse is as mystifying as superposition, sparking heated debates among physicists about the dynamics of observation and the nature of reality.

The phenomenon is not just theoretical. Consider the double-slit experiment: when photons are fired at a barrier with two slits, they create an interference pattern typical of waves, not individual particles. But, if we try to sneak a peek at which slit a photon passes through, the wave-like behavior vanishes, and they behave like particles. It's as if these fundamental bits of matter are actors that only "perform" their particle roles when observed.

Indeed, the implications of quantum superposition and measurement extend deeply into our understanding of the universe. Let me guide you through a fascinating realm where probability reigns supreme, and observation is an act of creation.

As we enter this quantum theater, the actors—particles—exist in a probabilistic limbo described by a wave function. It's critical to realize that these functions do not describe a fuzzy cloud of where a particle might be, but rather the probabilities of finding the particle in a particular state. This subtle difference is the heart of quantum mechanics. It's not that particles don't have precise locations and properties before we measure them; it's that these concepts don't have clear definitions outside of measurements.

Now, you might ask: Does this mean that reality doesn't exist without observation? Not quite. It's more accurate to say that the act of measurement forces nature's hand to choose a reality from a smorgasbord of possibilities—which isn't the same as creating it. The universe, it appears, functions with a set of probabilistic rules until measurements carve out a slice of determined existence.

But what does measurement truly entail? Is it the conscious observer that brings reality into focus? While it's tempting to place ourselves at the center of this quantum play, the truth is that any interaction with a quantum system can trigger wave function collapse. It might be a photon striking a detector, or a molecule bumping into another. In this way, the world is constantly observing itself, with countless interactions causing wave functions to collapse everywhere, all the time.

The Quantum Zeno Effect adds another twist to the tale. If a system is observed continuously, or rapidly enough, it's prevented from evolving.

Picture watching a pot, willing it never to boil—it's the quantum version of that adage, but with actual scientific backing. This perplexing effect implies that by merely watching, we can freeze the evolution of quantum states—a concept that shakes the foundations of causality.

So, how did this counterintuitive phenomenon find its way into reality? To tackle this, we traverse decades of experimental evidence gathered from particle accelerators, quantum optics labs, and even tabletop experiments with photons and atoms. The triumph of quantum theory is its unparalleled precision in prediction and its experimentally verifiable outcomes, underpinning modern technology from semiconductors to MRI machines.

But, as big a leap as it was, embracing superposition and wave function collapse is necessary for more than theoretical satisfaction—it has practical repercussions. From the microchips that power your smartphone to the lasers scanning your groceries, quantum mechanics is silently at work, exploiting superposition to perform wonders at scales that once seemed inconsequential to our everyday lives.

In the universe of quantum mechanics, mastery doesn't come through conquering these concepts but through partnership. We embrace the probabilistic elegance of superposition, the subtlety of measurement, and the counterintuitive implications as we witness our quantum world unfold in bewildering and fascinating ways. And as we stand at the precipice of the quantum revolution, with computation, communication, and sensing capabilities poised for transformation, the mastery of these principles is not just for intellectual gratification—it's a gateway to participating in the next wave of technological evolution.

In conclusion, quantum superposition and measurement challenge and enrich our understanding of the physical world. Reconciling these principles with our macroscopic intuition may never be comfortable, but the journey promises to be as thrilling as any exploration humanity has ever undertaken. In the seemingly capricious realm of quantum mechanics, the act of observation wields the power to sculpt reality, leading us to question what it means to understand the very nature of the universe. Embrace the quantum dance of particles, and let curiosity guide you through the maze of probabilistic existence, as we edge closer to uncovering the ultimate truths woven into the cosmic tapestry.

Quantum Fluctuations and Cosmic Inflation

Welcome to the mesmerizing world of quantum mechanics, where intuition steps aside and wonder takes the reins. Quantum Mechanics in Technology—this realm isn't just a theory locked away in academic papers; it breathes life into the extraordinary devices and systems you interact with daily. Let's explore how quantum principles are transforming the tapestry of modern technology, threading into the very fabric of our lives.

Imagine a world where communication is so secure that eavesdropping becomes a relic of the past—a world where ultra-precise sensors detect the earliest whispers of diseases and powerful computers solve complex problems in the blink of an eye. This isn't the plot of a science fiction novel. It's the future being woven by the loom of quantum mechanics.

Quantum Technology: The Revolution under Our Noses

Quantum technology isn't just a glimpse into the future; it's already at our fingertips. It has, quite literally, given us a whole new way to see the world through quantum imaging technologies. These technologies extend beyond the limitations of classical optics, giving us the ability to capture images with light that has never actually interacted with the object being imaged—think of it as a kind of quantum ghost imaging. It's as if light can whisper secrets about the shape and density of objects without ever touching them.

This staggering capability is not only a marvel for scientific exploration but also holds immense promise for medical imaging. Imagine the possibilities: safe, non-invasive methods for peering inside the human body with unprecedented detail. The impact on early disease detection and diagnosis could literally save lives.

Bridging the Gap: From Quantum Phenomena to Practical Devices

The bridge between esoteric quantum phenomena and practical devices is built on the sturdy piers of creativity and ingenuity. Every quantum application we envisage requires meticulous engineering and a profound understanding of the quantum landscape to transform theory into reality. For example, quantum cryptography harnesses the inherent weirdness of quantum states to create a system so secure, it's virtually tamper-proof. That's quantum key distribution (QKD) in action, my friends—a science so secure that it's predicated on the laws of physics itself.

In the realm of quantum computing, we encounter qubits—the fundamental building blocks fashioned out of delicate states like superposition and entanglement. These qubits hold the key to unleashing computational powers unfathomable with our classical bits. Take the infamous traveling salesman problem; even for a perfect classical computer, finding the most efficient route between a vast number of cities would take millennia. A quantum computer, however, with its ability to process a multitude of possibilities simultaneously, could potentially crack this problem over a lunch break. This revolutionary speedup isn't just impressive, it could change the very nature of work in logistics, finance, and numerous other sectors.

The impending quantum leap is not without its hurdles, however. Qubits are temperamental dancers, easily thrown off by their environment—a phenomenon known as decoherence. Engineers and physicists worldwide are racing to find ways to stabilize these quantum states and scale up quantum systems. Quantum error correction and fault-tolerant designs are at the forefront of this endeavor; they're the guardians ensuring our quantum computers don't falter at the slightest nudge.

Quantum Metrology: Keeping the World Precisely in Check

Enter another vanguard of quantum technology—quantum metrology. Using principles like quantum entanglement, scientists can measure physical quantities with a precision that makes traditional instruments look like blunt tools. Atomic clocks, those paragons of precision, employ entangled atoms to keep time so accurately that they won't slip by even a second over millions of years. Navigation, telecommunications, and global financial systems all lean on this quantum-informed precision as their invisible backbone.

Quantum Materials: The Subtle Art of Manipulation

The art of manipulating materials at the quantum scale is crafting new substances with properties right out of a dream. These materials may be superconductors that carry electricity without resistance, or perhaps metamaterials that bend light in impossible ways. Transparent aluminum, insulating magnets, or materials that remember their shape—all these once belonged to the realms of science fiction but are now being coaxed into reality by quantum artisans.

Quantum-enhanced imaging and sensing – Unveiling the Invisible

Quantum-enhanced imaging employs the brilliance of entangled photons, drilling down to precision unseen with classical light. These methods, bubbling up to the surface of mainstream applications, could very well revolutionize how we conduct surveillance, navigate autonomous vehicles, and perform geological surveys.

Quantum sensing, on the other hand, takes advantage of the extreme sensitivity of quantum systems to detect minuscule changes in magnetic fields, gravity, and even temperature. Navigational systems might soon be freed from reliance on GPS, operating instead on quantum compasses that tap into the Earth's magnetic field. These sensors could even lead to a greater comprehension of volcanic activity, potentially predicting eruptions with unprecedented accuracy and saving countless lives.

Quantum Technologies: Nurturing the Seeds of Tomorrow

While we marvel at the quantum technologies of today, it's the seeds of tomorrow that stir the imagination. Quantum internet, with its promise of unbreakable security, could redefine data privacy, transforming sectors like law, defense, and intelligence. The development of quantum batteries, with their ability to charge instantaneously and hold vast amounts of power, could greatly benefit renewable energy storage and the automotive industry.

These advancements, arriving with the subtle swiftness of quantum particles themselves, remind us of one crucial fact: the future is quantum, and it's already unfolding. The interplay of these quantum technologies will craft a world abundant with opportunity and teeming with solutions to some of our most challenging puzzles. As we continue to unfurl the quantum tapestry, there's no telling the marvels that await.

Quantum mechanics in technology is not merely a chapter in a book; it's a living, evolving saga of human innovation. It's the ingenuity of quantum mechanics made manifest in the devices and systems that enrich our lives. We stand at the cusp of a new era where fundamental understandings yield practical revolutions. And, fellow travelers on this quantum journey, you're now equipped with a clear-eyed view of just how the invisible, probabilistic nature of quantum mechanics is crystallizing into a future full of tangible, transformative wonders.

IMPLICATIONS FOR UNDERSTANDING THE COSMOS

In our bustling, vibrant universe, the invisible and the minuscule hold sway over the grand and the observable. Quantum mechanics—a realm teeming with peculiar wonders—enters your everyday life, not as a silent intruder but as a brilliant enabler. Today, we'll delve into how this enchanting science has paved the way for two amazing innovations: quantum sensors and imaging technologies, instruments so sensitive and precise, they're like the Superheroes of Sensing.

Let me introduce you to **quantum sensors**—think of them as the eyes and ears attuned to the softest whispers and lightest shadows of the physical world. Traditional sensors, those faithful servants of measurement, have served us admirably, but they have their limits. Quantum sensors, on the other hand, play by the bewildering rules of quantum mechanics, exploiting phenomena such as superposition and entanglement to detect changes in the environment with astonishing accuracy.

Imagine trying to find a single needle not in a haystack, but in ten hay fields. Seemingly insurmountable, right? But this is where quantum sensors perform their magic, sifting through countless signals to identify the one that matters. They're employed in precision timing, gyroscopes, and accelerometers, making them inseparable from the technologies we rely on, such as GPS systems that guide our planes, ships, and even our daily commute.

These quantum advancements extend beyond navigational aids; they encompass fields like medicine, where they tenaciously monitor magnetic fields in MRI machines, giving us clearer windows into the complexities of the human body. But the real beauty lies in their subtlety—they can detect changes in gravity with such finesse that they could spot mineral deposits deep within Earth's crust or discern volcanic activity before it becomes a threat.

Let's shift our focus to **quantum imaging technologies**. Here, quantum mechanics gives us a lens to peer into realms where classical cameras falter. Think of it as endowing us with 'X-ray vision' to gaze through walls of noise and obscurity. One stellar example is the 'quantum radar,' an upcoming technology that could detect objects with far greater stealth than its conventional counterparts—it's akin to detecting the silent flutter of a butterfly's wings amidst a raging tempest.

But the crowning jewel of quantum imaging is perhaps the mind-boggling concept of 'ghost imaging,' where an image can be captured using light that has never actually interacted with the object being imaged. Yes, you heard that right! It's like painting a portrait of a person by only observing the shadow they cast. This profound ability arises from the entanglement of particles—when two particles are interlinked, information about one instantly informs us about the other, no matter the distance.

Applications of these technologies aren't confined to the pages of science fiction or the isolation of physics laboratories—they're increasingly finding their place in security, where detecting concealed weapons becomes less an act of intrusion and more a discreet confirmation. They help archaeologists to unveil secrets of the past without so much as touching a relic. This tool, this extension of our senses, could redefine photography, under the sea or across the stars, by capturing the nuances of light that current technologies simply gloss over.

And let's not overlook the significance of quantum sensors in environmental monitoring and climate change. They become our watchful guardians, detecting minute atmospheric changes or pollutants at unprecedented levels. They have the potential to analyze soil composition in agriculture, oversee forest health, or quietly sit aboard satellites, contemplating the thinning of our precious ozone layer.

Now, at this juncture, you may wonder, "Why aren't quantum sensors and imaging devices ubiquitous?" The path to integrating such quantum feats into the fabric of daily technology is strewn with challenges. Quantum states are delicate; a mere glance from the environment can disrupt their coherence. This vulnerability, known as 'decoherence,' is the dragon that scientists must continually outmaneuver.

But advancements in shielding these quantum systems from interference have seen a vigorous stride forward. We've leaped from theory to early-stage applications in record time, and with each passing day, the technology grows more robust, marrying the fragility of quantum effects with the sturdiness of practical design.

The future is bright—or perhaps it is more fitting to say 'superpositioned'—for quantum sensors and imaging technologies. Their path is still one of potential, sprinkled with hurdles, but it's a journey worth taking. Soon, they'll not just be a nifty tool for specialists, they'll inform the way we interact with the world, enrich our understanding of it, and yes, even protect it. They are the silent affirmers of quantum mechanics' profound role in the fabric of reality—whispering the universe's secrets to those willing to listen, to those who dare to peer into its quantum heart.

In the end, the takeaway is clear: While the quirks of quantum mechanics might seem locked in the esoteric halls of science, they are, in truth, vibrantly active in our daily lives. From steering the GPS in your car to unearthing archaeological treasures undisturbed, quantum sensors, and imaging technologies are but glimpses of a world transformed by the quantum revolution—a world where the peculiar and the practical dance together, entwined in a symphony of technological marvel. And as we stand on the brink of this new era, remember: the quantum realm may be minute, but its impact on our everyday lives is anything but small.

CHAPTER 7: PRACTICAL APPLICATIONS OF QUANTUM MECHANICS

7.1 QUANTUM MECHANICS IN EVERYDAY LIFE

QUANTUM SENSORS AND IMAGING TECHNOLOGIES

As we venture through the garden of quantum mechanics, nourished by the understanding of its principles and applications, we arrive at a particularly alluring blossom—the Quantum Internet and Secure Communication. It's here where we see quantum theory entwine with information technology, sprouting an infrastructure poised to revolutionize the way we think about and handle secure communication.

Imagine an Internet that can send information with such security that the laws of physics protect it from eavesdroppers. This isn't a far-fetched dream but a burgeoning reality known as the Quantum Internet. It's a network that utilizes the peculiar properties of quantum bits—or qubits—for the transmission of information, significantly enhancing the security and speed of our communications.

The traditional Internet relies on digital bits—those **1**s and **0**s—to encode information. While encryption has come a long way in safeguarding our data, it remains vulnerable to advances in computing power. Given enough time and resources, encoded messages can be decrypted. But quantum encryption, through a mechanism known as quantum key distribution (QKD), defies this vulnerability. Here, quantum properties such as superposition and entanglement ensure that any interference by a third party is immediately detectable.

With QKD, two parties share a secret key encoded in the states of qubits—in this context, typically photons. Thanks to Heisenberg's uncertainty principle, if an eavesdropper tries to measure these qubits, their state inevitably changes. Much like a perfectly rigged alarm system, any intrusion is noted, and the key becomes useless, alerting the legitimate communicators to the breach. They may then discard the compromised key and initiate the distribution of a fresh one. In this quantum paradigm, the security of communication doesn't rely on the complexity of the encryption but rather on the fundamental principles of quantum mechanics.

The Quantum Internet goes beyond secure communication; it paves the way for a vast ecosystem of quantum technologies. Networks of quantum computers could share processing power and solve problems too complex for standalone machines. Moreover, by linking advanced telescopes with quantum networks, we could undertake measurements of the universe with unprecedented precision, ultimately untangling the cosmic mysteries that have long captivated the human imagination.

But how close are we to realizing this vision of a Quantum Internet? Researchers across the globe are working tirelessly to overcome the technical hurdles staring us in the face. The most significant of these challenges is quantum decoherence—the tendency for the delicate state of qubits to degrade over time and distance. However, the development of quantum repeaters, which act as amplifiers and restorers of quantum states, is helping bridge larger distances without loss of information fidelity.

Another promising technology essential for the Quantum Internet is the quantum satellite. By using space as a conduit for quantum signals, we sidestep terrestrial limitations such as signal loss through optical fibers. China has surged ahead in this space with the launch of Micius, a satellite specifically designed for quantum communication experiments. It has successfully demonstrated QKD over distances of thousands of kilometers, a giant leap toward a global Quantum Internet.

As exciting as these technological strides are, the Quantum Internet is more than a mesh of impressive gadgets. —it's a leap towards a future resonant with opportunities. Consider the world of finance—an industry where secure communication is paramount. Quantum secure networks could shield transactions from cyber threats, while quantum computing could optimize trading strategies and manage risk with algorithms that deftly handle market complexities.

Moreover, the fusion of quantum communication with healthcare can lead to unprecedented safeguards of patient privacy, as medical data can be transmitted securely over quantum channels. This ensures the confidentiality of personal health records, a concern in the digital age that's as vital as the very health care it seeks to protect.

Yet, as we traverse this revolutionary path, we ought to pause and reflect on the ethical and societal ramifications of the Quantum Internet. Its advent will make certain cryptographic systems obsolete, a transition that must be managed with the utmost care to prevent data breaches and potential chaos. We must also consider those who might be left behind—for every technological leap forward, we risk widening the digital divide.

Universities and private companies are indeed heeding this call, fostering initiatives that aim to educate and prepare the workforce of the future. Quantum literacy is becoming an increasingly valuable commodity, one that can unlock doors to opportunity and advancement. Furthermore, an international consortium of researchers and engineers is laying the groundwork for quantum communication standards, ensuring that the Quantum Internet becomes a shared resource, much like its traditional predecessor.

As we look ahead, we must balance our exhilaration for the advancements on the horizon with the responsibility of stewardship. We are, after all, the architects of this nascent quantum world, and our choices will resonate through the echelons of tomorrow. By championing accessibility, ethics, and open collaboration, we position ourselves to harness the Quantum Internet not just as a feat of science but as a testament to our collective spirit of progress and shared human experience.

In wrapping up our foray into the Quantum Internet and Secure Communication, remember that the most profound impact of quantum principles lies not in the technology itself, but in the potential, it unlocks: an interconnected society shielded by the immutable laws of the quantum realm, a promise of a secure, informed, and collaborative future. As quantum pioneers, when we expand the frontiers of our knowledge and technology, we weave a reality once captive to the imagination into the tangible fabric of daily life. Let the quantum adventure continue, with the veil of uncertainty lifted, revealing a universe of possibilities waiting to bloom.

QUANTUM METROLOGY AND PRECISION MEASUREMENT

In the fabric of our daily lives, quantum mechanics may seem like a distant twinkle of incomprehensibility, yet it is steadily stitching itself into the very essence of our existence. Perhaps nowhere is this more pronounced than in the arena of medical diagnostics, where quantum sensors are emerging as transformative instruments. These devices, though microscopic in form, offer monumental strides in the precision and efficiency of medical detection.

Imagine a world where the early detection of diseases is not only possible but routine, and the monitoring of health conditions is as effortless as holding a smartphone. This is the promise held within quantum sensors, leveraging the subtle fluctuations of quantum mechanics to probe the biological nuances of the human body with unprecedented resolution. They work by exploiting the sensitivity of quantum states to external changes, such as magnetic fields, temperature, or pressure, allowing them to detect signs of disease at scales unimaginable with current technology.

Healthcare professionals are on the cusp of having tools that can diagnose conditions from a simple breath test, potentially revealing markers for diseases like cancer, diabetes, and various infectious pathogens. These quantum-enabled devices can identify specific molecules at concentrations as low as parts per billion, making the proverbial 'needle in a haystack' search a tangible reality. It's akin to finding the exact word one seeks in an entire library, by the virtue of it humming a different tune.

Quantum sensors tap into this potential by employing technologies such as Nitrogen-Vacancy centers in diamonds, which are capable of sensing minute changes in magnetic fields produced by neurons firing in the brain. This can lead to groundbreaking advances in neuroscience, where the mysteries of the brain's circuitry and disorders can be mapped with an almost GPS-like precision.

Doing away with the bulky and intrusive machinery of today, future diagnostics may utilize quantum sensor technology in the form of wearable devices. These 'quantum wearables' could constantly monitor a patient's vitals and biochemical signals, sending alerts at the first sign of an anomaly. The implications for managing chronic diseases are revolutionary — imagine a diabetic patient receiving real-time feedback on their glucose levels without a single prick of the finger.

Quantum sensors also hold the prospect of dramatically improving imaging techniques. Magnetic Resonance Imaging, or MRI, an indispensable tool in diagnostics, could undergo a renaissance with the integration of quantum sensors. Currently, MRI machines are giants — they're expensive, consume vast amounts of power, and can intimidate patients with their claustrophobia-inducing design. Quantum technology promises MRI machines that are smaller, more energy-efficient, and sensitive enough to provide detailed images at much lower magnetic field strengths, reducing both cost and the potentially discomforting experiences for patients.

In the context of precision medicine — where treatments are tailored to the individual genetic makeup of a patient — quantum sensors could play an instrumental role. By enabling the detection and analysis of the unique genetic code within each patient's cancer cells, healthcare providers can better target therapies to attack malignancies with less collateral damage to healthy cells.

But what stands between the current state of medical diagnostics and this future filled with quantum wonder is a series of challenges and hurdles befitting any path of true innovation. The journey is not without its share of bumps, but understanding the destination makes the effort undeniably worthwhile. The intricacy of quantum-based technologies demands rigorous experimentation, validation, and approval by medical regulatory bodies to ensure efficacy and safety. Calibration, too, is no small feat when dealing with the subtle and delicate nature of quantum states, which can be disturbed by the slightest environmental fluctuations.

Moreover, ushering in these technologies necessitates a paradigm shift not just in the clinical instrumentation, but also in the mindset of healthcare professionals and the structure of medical practice. It is a grand endeavor to prepare the medical community for the integration of quantum sensors, requiring education on these new technologies and their potential impact on patient care.

Morally, it is a responsibility to ensure the equitable distribution of such advancements, so the benefits of quantum diagnostics can reach all corners of society. The chasm of healthcare inequality must not be widened by the arrival of new technologies but instead bridged with conscious effort and policy.

A visionary approach to future healthcare implies recognizing the transformative power of quantum sensors. The interplay of minuscule particles and their interactions is not just philosophical musing but the scaffold upon which the health of billions may be improved. This realm, rooted in the quantum depth, fertilizes our earthly existence with applications that could herald a new dawn of medical diagnostics, sharpening our insights into the human body and elevating the standard of personal health.

Much like the early explorers who embarked upon uncertain seas, we stand at the cusp of a vast ocean of quantum possibilities. With advances in medical diagnostics on the horizon, the wonders of quantum mechanics not only touch upon our deepest scientific curiosities but firmly grasp the tangible and critical realm of our well-being. As we follow this inevitable surge towards a healthier society, let us do so with hope and anticipation for a future defined by the fusion of quantum mechanics and medicine, for a world where the smallest particles yield the largest impact on human health. And it is in this quantum dance of particles and possibilities that we find a path to not only a deeper understanding of the universe but also a profound connection to the very essence of life.

QUANTUM ENHANCED MATERIALS AND DEVICES

Imagine stepping into Wall Street, not armed with traditional computers, but with the enigmatic power of quantum computing. Your portfolio pulsing with algorithms that operate on the edge of reality, slicing through financial noise like a hot knife through butter. This might sound like science fiction, but it's the burgeoning reality of where quantum computing in finance and optimization could soon take us.

Quantum computing represents a paradigm shift from classical computing. Instead of bits, we have qubits; rather than binary certainty, we're dealing with probabilities and superposition. In the crucible of financial markets and optimization problems — arenas defined by complexity and the need for lightning-quick computations — quantum computers herald an era of unprecedented potential.

Quantum Computing in Finance

Finance is an industry that thrives on prediction and risk management. From pricing assets to managing risk, financial institutions constantly seek tools that provide a competitive edge. Traditional computers have carried the industry far, but they're brushing up against their limits, particularly in areas that require processing massive datasets and modeling complex systems, where the fabric of financial computations resounds with the probabilistic underpinnings of quantum mechanics.

Algorithmic Trading

In high-frequency trading, milliseconds can make the difference between profit and loss. Quantum computers are poised to revolutionize this space with their ability to process vast amounts of information almost instantaneously. Quantum algorithms could analyze market data, accounting for a multitude of variables, to execute trades at optimal prices, harnessing not only speed but also a level of depth in prediction that classical computers simply cannot match.

Portfolio Optimization

Crafting the perfect portfolio is akin to finding the Holy Grail in finance. It involves weighing numerous financial instruments against each other, considering correlations, volatilities, and expected returns. Quantum computing offers the ability to process the complex mathematical models underlying portfolio optimization much faster than classical computers, considering far more scenarios and variables. This capability could lead to the construction of more robust, cost-effective portfolios that could withstand market volatilities and deliver better returns.

Risk Management

Managing risk is paramount in finance. Whether it's credit risk, market risk, or operational risk, quantum computing can dissect these multifaceted problems with finesse. For example, the Monte Carlo method, a statistical technique used for modeling the probability of different outcomes in a process that cannot easily be predicted due to random variables, is computationally intensive. Quantum computing can perform these simulations in a fraction of the time, allowing financial institutions to understand and mitigate risks more dynamically and accurately than ever before.

Quantum Computing in Optimization

In the ongoing quest for efficiency and profitability, businesses continually face optimization problems. Whether it's logistics, supply chain management, or decision-making, these problems can be so complex that today's computers can't find the best solution within a reasonable time frame, if at all.

Supply Chain Management

Quantum computers can delve into the intricacies of supply chains with incredible precision. Consider the logistics of a global company with multiple warehouses and a multitude of transport routes. Classical computers can find good-enough solutions for managing these resources, but quantum computing has the potential to uncover the optimal solution, considering more variables and constraints, thereby reducing costs and improving service.

Energy and Resource Allocation

The energy sector can benefit from quantum optimization by efficiently allocating resources. This has immense implications for renewable energy management, where the supply is as variable as the weather. Quantum algorithms can calculate the optimal distribution of electricity from irregular sources, like wind and solar farms, to where it's needed most, maximizing efficiency and sustainability.

Strategic Decision-Making

One of the most exciting prospects for quantum computing lies in strategic decision-making. Imagine being able to model countless scenarios and outcomes for business decisions, playing out the complex interactions between various elements of the business environment, competitors, regulatory changes, and market forces. Quantum computers could offer insights by identifying the paths that maximize success potential, leading to decisions grounded not just in gut feel but in quantum-aided foresight.

Challenges and Considerations

As we gaze at the horizon of what quantum computing can bring to finance and optimization, we also need to shelter our optimism with the practicality of imminent challenges. Decoherence and qubit stability are still thorns in the side of quantum computing's full-scale deployment. Substantial investment and research are focused on these technical hurdles, inching ever closer to reliable, scalable quantum systems.

Then there's the question of accessibility. As with any revolutionary technology, the initial wave of quantum computing resources is likely only to be available to the most well-funded organizations. We must be cognizant of the potential for widening the gap between the 'quantum haves' and 'have-nots', and strive to democratize the advantages that this technology holds.

Furthermore, powerful as they may be, quantum computers won't render classical computing obsolete. The two will likely coexist, with quantum machines handling tasks that are suited to their unique capacities and classical machines managing the rest. Quantum algorithms are fundamentally different from their classical counterparts and will require a new breed of programmers and financiers fluent in quantum logic.

Conclusion

Quantum computing is not the final piece of the puzzle in finance and optimization; it's an entirely new puzzle waiting to be solved—an intricate blend of high-stakes finance and cutting-edge technology. While daunting, our exploration into the quantum realm is a testament to human ingenuity and our incessant drive to push the boundaries of what's possible.

As you close the pages of this book, remember the journey of learning quantum mechanics is one of patience, curiosity, and wonder. The principles you've learned here are the foundation upon which applications in finance and optimization rest—brick by quantum brick, we're building a future that once existed only in the confines of our wildest imaginations.

7.2 Future Trends and Emerging Technologies

Quantum Internet and Secure Communication

In the mesmerizing dance of subatomic particles, quantum entanglement plays a starring role, a dance where distance is a mere suggestion and intuition takes a backseat to the astounding reality of quantum mechanics. You see, quantum entanglement is not just a perplexing scientific concept; it's the underlying magic that unlocks an array of breathtaking technologies and revolutionizes our understanding of the universe.

Picture two particles, born from the same quantum event, akin to cosmic twins separated at birth. These particles, regardless of the miles or light-years between them, remain intertwined in a delicate ballet of shared existence. If you whisper a secret into the ear of one, the other, no matter how far away, instantaneously catches your hushed words. This is entanglement, and while Einstein famously rebuked it as "spooky action at a distance," it remains an undeniable pillar of the quantum world.

But before we venture further into the riddle of entanglement, let us ground ourselves in Bell's Theorem. Proposed by physicist John Bell in the **1960s**, this theorem challenged the classical view of the world, suggesting that either the universe is interconnected in ways we don't yet understand (non-locality), or the properties of particles are hidden away from our current perception (hidden variable theories). Subsequent experiments lent weighty support to non-locality, painting a picture of a universe far more interlinked than we ever thought possible.

Turning our gaze to the EPR Paradox, we unravel an intellectual thread that began with a challenge from Einstein, Podolsky, and Rosen. Their argument struck at the heart of quantum mechanics, questioning whether a quantum state is a true representation of reality if it cannot account for precise properties of particles without measurement. The paradox keenly illustrates the tension between quantum mechanics and our classical expectations—a tension that, to this day, prods at the bedrock of our understanding.

Our quantum adventure accelerates as we delve into the practical side of this phenomenon: quantum cryptography and secure communication. Imagine transmitting a message with the absolute certainty of privacy. Through entanglement, quantum cryptography allows just that—an unbreakable encryption not reliant on computational complexity but on the laws of physics themselves. Companies are already piloting quantum networks where eavesdropping becomes an act of futility; any intrusion irrevocably changes the state of the entangled particles, sounding an undeniable alarm.

Then we tiptoe into the realm of quantum teleportation. Not the beam-me-up-Scotty kind, but rather the transmission of quantum information—qubits—across a distance. It's a process that harnesses entangled pairs to transfer the quantum state of a particle to another, effectively "teleporting" its properties across space, leaving behind the original particle. This extraordinary process has profound implications for quantum computing and the relay of quantum information.

However, it is not without its challenges. The delicate state of entangled particles is prone to interference from their surrounding environment—a phenomenon known as decoherence. It's akin to trying to hear a whisper in a roaring stadium. This fragility renders the storage and management of quantum information a daunting task. Researchers are actively seeking ways to stave off decoherence and preserve the entangled state, ensuring that its potential can be fully realized in practical applications.

One might question, what does the future hold for entanglement research? Here, the air buzzes with excitement. Within this kaleidoscope of possibility, we find quantum networks—vast webs of entangled particles transmitting information across the globe. Imagine an internet that thrives on the currency of qubits, fueled by the speed and security of quantum mechanics. As the race to realize a quantum internet picks up pace, it will bring with it a dawn of faster, more secure communication that reshapes the foundational fabric of how we connect and share information.

Our journey through the enigmatic landscape of quantum entanglement comes to a close, but the path forward only stretches out further and wider. This is the beating heart of quantum mechanics in action, exemplified not just in secure communication and futuristic teleportation, but in the rhythm of the universe itself. One cannot help but feel an echo of that quantum interconnectedness within ourselves—a sense of awe that transcends the page and propels us into the vast unknown of the quantum frontier.

Let us remember, though, that amidst the swirl of probabilities and superpositions, quantum mechanics is an invitation—a call to widen the aperture of our minds. So, when faced with concepts that reach beyond our classical intuitions, let us greet them not with skepticism but with the curiosity and wonder they deserve. In the dance of the quantum world, it's this very spirit of exploration and the relentless pursuit of understanding that ensures the music will play on, leading us to realms once thought unreachable.

QUANTUM SENSORS FOR MEDICAL DIAGNOSTICS

Imagine, for a moment, that you're able to do something utterly astonishing—something so counterintuitive it bends your perception of reality. Picture a particle, a tiny speck of matter, and now see it spread out like a wave, rippling across an ocean of possibility. This isn't the stuff of science fiction; this is the reality in the quantum realm. This peculiar behavior where particles sometimes act like waves, and waves behave like particles, is the essence of wave-particle duality.

In the heart of every atom and in the photons of light that reach us from the sun, wave-particle duality is at play. It answers old-age queries like why electrons don't crash into the nucleus and how phenomena like interference patterns emerge. This duality was first postulated through the brilliant mind of Louis de Broglie, who suggested that if light, traditionally viewed as a wave, could be shown to have particle behaviors, then perhaps particles, such as electrons, might exhibit wave-like characteristics.

The cemented concept sprang from early **20**th century experiments. When shining a light on metal surfaces, surprisingly, it ejected electrons. This photoelectric effect, as it's known, couldn't be explained by the physics of the time, as waves didn't have the localized impact to knock out electrons from metals. The only way to explain it was to accept that light was acting as a particle—a photon with a quantized punch. Another phenomenon, the Compton scattering, reinforced this, showing light changing frequency after bouncing off electrons, as if particles were colliding.

These discoveries were groundbreaking, but it was de Broglie's hypothesis that tied everything together. Every particle has a wavelength, de Broglie proposed, and the smaller the particle, the more significant its wave nature.

This led to experiments that showed electrons creating interference patterns, which only waves were expected to do—unless, of course, those electrons were waves themselves when not being directly observed. And thus, the duality became a central tenet of quantum mechanics.

What can appear perplexing is that this duality is typically unobservable in our macroscopic world. You've never tossed a ball and watched it spread out into a wave and interfere with itself. But at the quantum scale, particles such as electrons do exactly that. They display a probability wave, spread across space and time, describing the likelihood of finding the electron in any particular location. In essence, before measurement, the particle is in a spread-out state, with a myriad of potentialities for where it might end up.

The interplay between the wave-like and particle-like properties has astonishing practical consequences. It's within the belly of every electronic device you use. Semiconductor technology, lasers, and even your smartphone's camera—these wouldn't exist without our grasp of wave-particle duality. Quantum tunneling, too, emerges from this principle. Where classical particles would face an insurmountable barrier, quantum particles can pass through as if the barrier was just a mirage. This tunneling is the magic behind the workings of the microprocessors and flash memory.

Another practical application is the electron microscope, capable of peering at the smallest of structures that light-based microscopes can't resolve. It uses the wave nature of electrons—focused into a beam—to illuminate samples. Since electrons have a much shorter wavelength compared to visible light, they can show us the fine details of viruses or the intricate structure of materials, pushing the boundaries of what we can see.

This duality is not just confined to electrons; it's a universal characteristic of all quantum particles. It implies that particles are not just discrete entities but also exist as waves of probability—until they are pinned down by a measurement. An electron can be a wave smeared across the cosmos or a point particle sitting on an atom; it depends on how you look at it. This idea might seem to be an esoteric quirk of nature—curious but distant from our everyday experience. Yet it is immensely practical, becoming the cornerstone for technologies that you use every day.

One of the marvels of wave-particle duality is its pure elucidation of nature's versatility. It is a feature that propels not only electron behavior but also the operations of all fundamental particles. Its implications for quantum computing are especially thrilling. As you delve deeper into understanding the quantum world, you'll uncover the beauty of wave functions and qubits—underpinnings of a future where computation soars beyond the limits of today's binary systems.

Lastly, let's consider the philosophical implications of wave-particle duality. It challenges our classical intuition of an independently existing, well-defined world. The quantum reality it describes is a more delicate fabric, contingent on measurements and full of probabilities. Embracing this duality allows you to imagine a world more intricate and dynamic than anything offered by classical physics—a world where the fundamental building blocks are always dancing between the definite and the indefinite.

In the grand canvas of quantum mechanics, wave-particle duality paints a startling yet coherent picture of reality that is fundamental to how we comprehend and interact with nature. As you ruminate on this profound phenomenon, know that it's just one piece of a much larger and even stranger quantum puzzle.

As we tread further into the landscape of quantum mechanics, it's beneficial to move with an open mind and a sense of wonder, for the principles we're unraveling aren't just intricate pieces of an academic discipline; they're the whispers of the universe itself, inviting us to understand the nature of reality at its most fundamental level.

QUANTUM COMPUTING IN FINANCE AND OPTIMIZATION

In the symphony of quantum mechanics, there are a few off-key notes that remain persistent challenges for scientists and engineers alike. One such hurdle is the duo of decoherence and noise, which are akin to the unwanted disturbances that can disrupt a delicate performance. At their core, these phenomena stem from the seemingly indomitable interaction of a quantum system with its environment—a tango dance, if you will, where the missteps of environment-induced vibrations can lead our quantum particles astray from their choreographed routine.

Decoherence: Picture, for a moment, a quantum particle as an expert dancer, moving with a purpose and grace that exhibits its characteristic superposition—where it can exist in multiple states, poised for any number of possible performances simultaneously. However, as the audience (the environment) gets involved, clapping hands and tapping feet, the dancer's moves become less certain. This is decoherence: the loss of this delicate quantum superposition due to environmental disturbance, causing the quantum information to seemingly dissipate like whispers in the wind.

Unlike the classical information inscribed in the pages of a book, quantum information encoded in the state of a particle is far more susceptible to the side glances of the universe. Each glance, each particle from the environment that interacts with our quantum states, is a nudge toward the classical world, coaxing our quantum dancer to choose a particular position, one definitive state, and abandon the quantum waltz.

Now turn your gaze to **quantum noise**, which can be thought of as the dissonant background chatter that muddies the signals, we so wish to discern clearly within a quantum system. It stems from a variety of unwanted fluctuations—thermal, electromagnetic, or even the touch of a cosmic ray—and reshapes the landscape of energy levels, timing, and probabilities so crucial for quantum activities to play out in concert.

Overcoming these twin trials, one might argue, is like seeking to perform the most intricate ballet in the middle of a bustling city intersection. Yet, if quantum mechanics is to make the leap from the chalkboard into the concrete jungle of everyday application, this is the dance we must master.

Research brigades have been marching toward various strategies aimed at outmaneuvering decoherence and minimizing noise. Some choose to confront them head-on, developing new materials with intrinsic properties that ward off the environment's peering eyes, akin to an invisibility cloak spun from the threads of human ingenuity. Others delve into what is known as 'quantum error correction,' a series of protocols that can detect the errant steps induced by decoherence and noise, whispering timely corrections to our quantum dancer.

These error-correcting codes are the equivalent of a tightrope walker's balancing pole, gracefully applied algorithms that counteract the wobbles before a tumble ensues. However, much like the need for reliable safety nets, quantum error correction demands redundancy, requiring large numbers of physical qubits to protect the sanctity of fewer, 'logical' qubits that hold our precious quantum information.

The aspiration to insulate our qubits from the incessant chatter of the universe has led to the quest for 'decoherence-free subspaces' – niches in the quantum landscape untouched by the cacophony of the outside world, where quantum states can flourish unaffected. Engineering these subspaces is akin to crafting a perfectly soundproof room in a house perched beside railway tracks.

Quantum engineers have rolled up their sleeves to face another daunting task: **taming entanglement**, the very phenomenon that provides quantum computing with its awe-inspiring parallel processing prowess but which is incredibly sensitive to decoherence. Imagine trying to manage a network of trapeze artists, all swinging in perfect harmony, while gusts of erratic wind threaten their delicate balance. Preserving entanglement among qubits over significant distances and timeframes is one of the state-of-the-art performance achievements that we're striving toward.

We also cannot neglect the role of **quantum control theory**, a sophisticated suite of mathematical strategies that choreographs the behavior of quantum systems with ever-growing precision. It aims to preempt the influence of noise and correct the system's trajectory as one might correct the course of a sailboat amidst gusty winds and churning waves. Advanced control schemes employ adaptive methods that learn and predict the impact of noise, guiding the quantum state through the storm with deft adjustments.

The advancement of quantum technologies has been propelled by the growing mastery of techniques such as **dynamical decoupling**, which could be considered the noise-canceling headphones of the quantum world. Periodic pulses are applied to the quantum system, effectively drowning out the environmental noise and allowing the qubits to retain their coherence for extended periods.

But let us not ignore the elephant in the room: the current **physical limitations** of our quantum apparatus. Constructing quantum systems with the fidelity required to mitigate noise and decoherence remains a monumental challenge. Every wire, every junction, every material in a quantum processor must be as flawless as we can muster, as even the smallest imperfection could cascade into quantum chaos.

Despite these steep mountain ranges traversed by decoherence and littered with noise, there's an air of optimism that winds through the quantum community. As we refine our tools and deepen our understanding of how the quantum world ticks, we learn to dance with, rather than against, the peculiar rhythms of decoherence and noise. The incentive pushing us forward is none other than the potential to revolutionize computing, communication, and beyond.

In conclusion, while decoherence and noise present intimidating barriers, they are not insurmountable. The field advances with every experimental triumph and each theoretical breakthrough, gradually paving the way for quantum systems robust enough to face the chaos of the real world. As we seek to cozy quantum mechanics onto the fabric of daily technology, we embrace these challenges as opportunities—a call to arms—to bring about the next wave of scientific and technological innovation. Just as the once elusive notion of flying has become a mundane reality, the day may come when quantum robustness is looked upon as yet another mountain conquered in the human pursuit of the extraordinary.

CHAPTER 8: CHALLENGES AND FUTURE DIRECTIONS

8.1 CURRENT CHALLENGES IN QUANTUM PHYSICS

OVERCOMING DECOHERENCE AND NOISE

In our quest to decipher the universe's most closely guarded secrets, we often find ourselves standing on the precipice of the unknown. Quantum Information Science and Quantum Networks form the vanguard of our voyage into this exciting future of discovery and technological innovation. Pioneers in the field eye these as nothing less than instruments that will redefine the way we interact with data and with each other.

Quantum Information Science (QIS) extends its roots deep into the heart of quantum mechanics. It is a domain where information is processed and communicated in accordance with the principles of quantum theory. This approach to information introduces fascinating possibilities that traditional binary-based systems can never hope to achieve. QIS is not merely a different way to do what we've always done; it's the bedrock for things we haven't even imagined yet.

Within the intricate dance of particles that exhibit both entanglement and superposition, QIS seeks to capitalize on these phenomena to revolutionize computation, communication, and encryption. Imagine a world where computers perform calculations at speeds that dwarf today's most powerful machines, or messages that are sent with a security so robust that they are impervious to interception or hacking. This is not just a theoretical playground; it's a future that's slowly crystallizing before our eyes.

At the heart of QIS are quantum bits, or qubits, the building blocks of quantum computing. Unlike classical bits, which are confined to being either **0** or **1**, qubits can exist in multiple states simultaneously thanks to superposition. This characteristic alone propels quantum computing to stratospheric heights of parallel processing power. It's as if a coin in quantum computation can spin indefinitely, landing on heads, tails, and a spectrum of states in between when measured.

Leveraging entangled qubits, quantum networks add another layer to this paradigm-shifting scenario. Once particles are entangled, the state of one instantly influences the other, regardless of the distance separating them – a property Albert Einstein famously referred to as "spooky action at a distance." Quantum networks harness this instantaneous connection to transmit information in ways that traditional networks can't parallel.

But beyond the allure of instantaneous communication lies the promise of unbreakable encryption. Quantum key distribution (QKD) utilizes the principles of quantum mechanics to create keys that any attempt at interception would irrevocably alter, signaling any breach in communication integrity. This isn't just an incremental step up in security; it's an absolute game-changer in a world where privacy and data protection are ever more crucial.

The potential of QIS stretches beyond raw computational power and impregnable communication. With quantum networks, we stand on the brink of forming a quantum internet, a network where quantum processors are linked across distances, sharing information and resources. This quantum web won't just be a faster version of our current internet; it will enable entirely new forms of distributed quantum computing, conjure up unprecedented algorithms, and foster collaboration that spans continents without delay.

Given the gravity of such advancements, it's no surprise that researchers and engineers face a gauntlet of obstacles. Quantum systems are fragile, easily disrupted by external environments – a challenge known as decoherence. Keeping qubits in their quantum state long enough to perform meaningful operations is one of the most significant challenges facing the field. Likewise, scalability poses another conundrum. Today's quantum processors with their limited number of qubits need to evolve into more complex networks that host thousands, if not millions, of interlinked qubits.

There's also the hurdle of quantum error correction. In a world where our current logic of error-checking doesn't apply, creating systems that can self-correct at the quantum level is paramount for reliable operation. Without such mechanisms in place, quantum information systems might fail to deliver on their theoretical promise.

The road ahead for quantum information science and quantum networks reflects both our loftiest aspirations and our most sobering technical challenges. The future holds a collaborative effort that spans disciplines, from physics to computer science, from engineering to cryptography. It is an undertaking that will require not just skilled minds but visionary thinkers who can look beyond the horizon and imagine the potential of what could be.

As we stand on this threshold of innovation, it behooves us to ponder not only the technical implications but also the societal impacts of QIS. Alongside the awe-inspiring capabilities of these technologies, we must be vigilant about ethical considerations and ensure that the quantum leap in connectivity and computing power benefits all of humanity.

It is an exhilarating time to be alive for those with a passion for quantum mechanics and the transformative potential it holds. The foundation laid by QIS and quantum networks won't just alter the trajectory of technology; it will redefine the way we comprehend and interact with the fabric of reality itself. To the software engineers, research scientists, and the avidly curious who have joined me on this journey through the quantum realm, the future beckons with a promise of mysteries revealed and capabilities unlocked. Let's continue to explore with unyielding curiosity and a shared vision of a world increasingly empowered by the quantum revolution.

SCALING QUANTUM SYSTEMS FOR PRACTICAL APPLICATIONS

As we pivot our gaze to the horizon, the vista of quantum computing and simulation materializes — a landscape brimming with potential and fraught with the adventure of the unknown. The journey of quantum computing thus far has been like scaling a towering peak; with each hard-won step upward, the summit nonetheless remains shrouded in mist, teasing with promises of revelations yet to come.

Let's take a moment to consider the blueprint of classical computers. In these machines, a binary symphony performs — bit by bit — the accumulation of ones and zeros into elaborate architectures of information. However, these bits are bound by their binary nature; they are the steadfast sentinels of an ordered world, representing 'on' or 'off', 'true' or 'false', with unwavering fidelity. Quantum computing, by contrast, heralds a paradigmatic shift. It invites us into a realm where the quantum bit, or qubit, defies such binary constraints through the enchantment of superposition, both 'on' and 'off' simultaneously, until observed.

This inherent ability of qubits to hold multiple states concurrently gifts quantum computers with powers exorbitantly beyond the wildest dreams of their classical counterparts. Imagine a library where, instead of reading one book at a time, you could absorb every volume's wisdom in a single, ephemeral moment. Such is the might of quantum computation; where classical computers trudge through algorithms step by linear step, their quantum brethren survey a multitude of paths in parallel, collapsing time and complexity in their wake.

One of the most breathtaking arenas where quantum computing is set to galvanize change is in the field of simulation. Here, the natural language of quantum systems becomes an intuitive script to simulate the quantum world itself. Molecules, materials, and the mysterious interactions of quantum phenomena emerge as lucid through the quantum lens. Chemists, yearning to uncover new medicines, stand to leap forward as simulations elucidate potential drug interactions before a single test tube has been dirtied. Materials scientists, driven by the quest for novel substances, may find their holy grails in the sanctum of simulated spaces.

But it's not only the small-scale world that unfurls under the scrutiny of quantum simulation; the vast expanse of the cosmos, too, submits its secrets. Dark matter, that elusive scaffold of the universe, or the conundrums of black hole information paradoxes, once deemed inscrutable, inch toward the light of understanding. Quantum simulations promise to be the Rosetta Stone, translating cosmic whispers into a vocabulary accessible to our hungry minds.

Yet such powers come at a price. Quantum decoherence — the nemesis of qubit stability — remains the dragon at the gates, snuffing out superpositions with the harsh breath of interaction with the environment. The quest for quantum error correction looms large, with researchers actively weaving the spells of fault-tolerant quantum computing. Efforts multiply daily, winding protectively around the fragile heart of quantum computation.

The challenge of scaling, too, must be vanquished. Quantum processors of sufficient qubit count to tackle the meatier problems are still in gestation. Even with our best efforts, we are toddlers taking tentative steps; our most sophisticated quantum processors today are but playthings compared to the colossi we dream to build.

But bear heart, for advancements march on. The realm of topological quantum computing whispers of qubits that are not perturbed by the common scuffles of the quantum world. Embracing exotic states of matter, these tireless researchers craft qubits entwined in an intricate dance, where any attempt by decoherence to cut in is deftly parried.

The interplay between software and hardware shines a beacon to the future, too. Quantum algorithms spiral in sophistication — from Shor's algorithm, a harbinger of a cryptographical revolution, to Grover's algorithm, promising a search that surges like lightning. Innovators pen software to pivot seamlessly with the flux of quantum hardware, scripting the choreography of a dance most intricate and beautiful.

Let us not forget quantum networks, which tingle with the potential of a world intimately connected through entangled qubits. These networks, lacing cities, countries, and potentially the globe, would be a new vein of communication and computation. A quantum internet, securely fortified by the strange sentinels of entanglement, stirs the imagination with scenes of instantaneity and impregnability.

While the practicalities of implementation pace like caged beasts — the theorized qubit enormity, the elaborate orchestration of coherence, the spectral hand of error correction — we are, inch by inch, drawing the dream closer to reality. In recent triumphs, we've witnessed discrete quantum algorithms execute with precision, a testament to progress, each small victory stoking the fires of potential.

The pursuit of quantum advantage, that tipping point where quantum computers transcend classical performance for practical applications, ticks closer. It may well begin with a bespoke task, a particular algorithm's affinity with a quantum processor's strengths. But it will burgeon, blossoming into a reality where quantum simulation and computing become the cornerstones of industries, the forges of economies, the workshops of scientific revelation.

In the grand tapestry of technological growth, quantum computing and simulation stand as golden threads, promising to weave patterns of unimagined complexity and beauty. As the shroud lifts from the peaks we strive to summit, the vistas revealed will undoubtedly transform our grasp of the natural world, and with it, the substance of our reality.

Addressing Ethical and Societal Implications

In our exploration of the quantum realm, one of the most mind-boggling and yet profoundly impactful concepts we encounter is entanglement. It's a phenomenon that Albert Einstein famously referred to as "spooky action at a distance," and it has since intrigued and puzzled physicists and philosophers alike. But what does this really mean for us—in our daily lives and as we peer into the vast unknown of cosmic mysteries?

Let's start with the basics. Quantum entanglement occurs when pairs or groups of particles interact in ways such that the quantum state of each particle cannot be described independently of the state of the others, even when the particles are separated by large distances. To put it poetically, entangled particles are like star-crossed lovers bound by an invisible string that connects their fates, regardless of the space between them.

The implications of this are profound. The entanglement challenges our very notions of space and time and defies the idea that for an object to affect another, it must be in close proximity. This phenomenon has a ripple effect, reaching into the depths of space. Imagine, if you will, the mystifying black holes—objects so dense that not even light can escape their gravitational pull. The information that falls into a black hole, one might think, is lost forever. But if that information is quantum mechanically entangled with particles on the outside, could it somehow escape the clutches of the black hole's event horizon? This is one of the mind-twisting puzzles that quantum entanglement presents in astrophysics.

In the cosmos, entanglement might be a thread in the fabric of space-time itself, suggesting a universe where connections transcend the classical laws that we thought governed space and time. Some physicists hypothesize that quantum entanglement could be a cornerstone for a theory of quantum gravity, a theory that has eluded scientists, which aims to unify the seemingly incompatible theories of quantum mechanics and general relativity.

Quantum entanglement doesn't just stop at the theoretical. It has the potential to transform technology and computing, leading us into a new era. It could enable quantum computers that perform calculations at speeds unimaginable with today's computers. Critical problems in drug discovery, materials science, and artificial intelligence could be solved in the blink of an eye.

But the journey is not without its challenges. Engineering entangled states in the lab is a delicate dance, and maintaining these states long enough to do something useful with them is even more delicate. This is because entanglement is fragile—disturbances from the environment, known as decoherence, can easily disrupt this special quantum state, breaking the entanglement. Overcoming these disturbances is a major hurdle for scientists working to harness the full power of quantum entanglement in practical applications.

And as we think about the future of quantum technologies, let's not forget the ethical and societal implications. Quantum cryptography promises secure communication that can't be hacked by conventional means. But what happens when malicious actors gain access to this same power? The balance of security and privacy is likely to be impacted in ways we have yet to fully imagine.

Let's dive a bit deeper into the ethical considerations. As quantum entanglement and the associated technologies mature, they will influence sectors well beyond the realm of computer science and physics, affecting everything from financial markets to national security. Questions arise: Who controls access to such powerful technology? How do we ensure that it's used for the benefit of humanity and not for destructive purposes?

Above and beyond its immediate applications, quantum entanglement compels us to confront the fundamental mysteries of the universe head-on. For instance, could entanglement be the missing piece in our understanding of the fabric of reality? Do we need to rethink the very nature of matter, space, and time?

Entanglement may become a sort of cosmic glue, offering insights into what tethers the universe together at the most fundamental level. Perhaps, entangled particles scattered across the universe are akin to a vast network—a quantum web interweaving all matter and energy, a basis for a new understanding of the interconnectedness of everything.

Ultimately, our journey through quantum entanglement is not just about untangling the technicalities of particles and probabilities—it's about interlacing the profound with the practical, weaving esoteric theory into the tapestry of everyday technology. It's where the seemingly impossible becomes possible, where the otherworldly becomes palpably present.

As we inch closer to a new age of quantum technologies, we also progress toward a higher plane of understanding. Each step forward, each entangled particle studied, and each quantum state preserved brings us closer to the realization that the universe is a far more interconnected and mysterious place than we ever imagined.

Let us embrace the strangeness, for it is in the peculiar and perplexing that progress often hides. Let us be guided by curiosity and inspired by the promise of what lies just beyond our current grasp. The quantum entanglement is but a keyhole view into the vast potential that awaits us as we unlock the secrets of the quantum realm and harness its power to redefine the boundaries of possibility.

In all the vastness of our universe, within the minutiae of particles dancing to the quantum tune, lies the opportunity for a greater comprehension of the cosmos and our place within it. It is here, in the heart of entanglement, where the next chapter of our cosmic story is waiting to be written. And so, with our quantum key in hand, we step forward—eager to turn the lock and swing wide the doors to a future replete with quantum wonders.

8.2 Future Directions in Quantum Research

Quantum Information Science and Quantum Networks

Embarking on a journey through the quantum realm, we often encounter a conundrum that seems to tease the limits of our understanding: how can something be in multiple states at once, and only settle into one upon observation? It feels as if the quantum world plays a game of hide-and-seek with reality, revealing itself only under the scrutinous eye of the observer. This concept, my friends, is aptly named quantum superposition.

If we conjure the image of Schrödinger's mischievous cat, simultaneously alive and dead within its unopened chamber, we find ourselves face-to-face with one of the most bewildering and beautiful paradoxes of quantum mechanics. The very essence of this thought experiment demonstrates the principle of superposition. Inside the box, before we peek, the cat's fate hovers in a delicate balance of life and death—quantum states piled atop one another like a deck of cosmic cards waiting to be pulled. It's only when we flip the latch, draw back the curtain, that the probabilities collapse into a single, definite state.

You may wonder how this could possibly mirror reality. Yet the universe, on its most fundamental level, is stitched together by such enigmas. Atoms and particles often exist in multiple states simultaneously—a single electron may travel through two paths at the same time when not observed, much like how a wave spreads across the surface of a lake, touching many pebbles at once. This is superposition in action, and the implications are profound. It means particles can interact with themselves, interfere with their own paths, and exhibit behaviors beyond any classical imagination.

The phenomenon of superposition isn't merely a philosophical dalliance; it forms the backbone of technological advancements. Take quantum computing, where information is stored in quantum bits, or qubits. Unlike traditional bits that hold a value of either **0** or **1**, qubits, thanks to superposition, can embody both values at once. This allows quantum computers to process an exponentially larger amount of information, a mosaic of probabilities, all at the same instance, promising unprecedented speeds for certain computations.

Intriguingly, superposition also whispers the promise of quantum teleportation. No, not the making-novels-of-our-physiques kind, but the transfer of a quantum state from one particle to another over a distance, without the physical movement of the particles themselves. It relies on our understanding of superposition and entanglement (a peculiar sort of quantum bond we'll explore later), portending a future where information may travel in ways that render our current communication technology positively quaint.

But let's address the elephant in the room—measurement. The act that lifts the curtain, forces Mother Nature's dice roll, if you will. The quintessential quandary of measurement in quantum mechanics lies in how the act of observing alters the observed. What does this mean? When we deploy a measuring device into the quantum soup, it interacts with a particle in superposition, instantly nudging it to adopt one of its possible states.

It's as if the universe operates on a form of quantum etiquette, where particles only pick an identity when politely asked by an observer. This is not to say consciousness creates reality (a popular misinterpretation); rather, it's the interaction—an unavertable physical process—that collapses the wave function into a definite outcome.

These concepts extend further into philosophical domains, asking whether reality itself is malleable or predetermined. The measurement conundrum invokes the Quantum Zeno Effect, imagine freezing a particle in a state just by measuring it frequently. Through endless observation, one stops the particle from evolving into a different state, a paradoxical twist that locks the quantum system in time.

As we peer into the heart of quantum mechanics, we uncover a nature that is both grand and granular. Every technological leap forward—be it in quantum cryptography, precision measurement, or materials science—is a testament to our growing understanding of superposition and measurement.

As you digest these notions, remember that such explorations are not solely the territory of seasoned physicists or mathematicians. Aspiring software engineers and curious minds alike are very much a part of this narrative. Quantum mechanics opens an emporium of possibilities, where the pursuit of the infinitely small empowers the grandest of dreams. The practical implications of superposition and measurement extend far beyond the walls of academic institutions; they are already reshaping the world we live in, revolutionizing the way we interact with the deepest truths of nature.

Be encouraged, my fellow quantum enthusiasts, for in this intricate dance of subatomic particles, we find not only answers but also the quintessential questions that push humanity forward. Through persistence, curiosity, and the occasional leap into the unknown, quantum mechanics becomes not just a discipline to study but a realm to experience and transform.

As we press on this journey through quantum mechanics, let's harness the beauty of superposition, the subtlety of measurement, and the transformative power they collectively wield. With every chapter, not just in this book, but in the pages of our exploration of the universe, let's aspire to be both the observer and the architect, unraveling and weaving the very fabric of reality.

ADVANCEMENTS IN QUANTUM COMPUTING AND SIMULATION

Imagine you're holding a coin—one side heads, the other tails. Now give it a flip, and let it land concealed in your palm. Without looking, can you tell which side is facing up? Of course not. But quantum mechanics offers a stranger perspective: until you observe the result, the coin is in a peculiar state where it's both heads *and* tails simultaneously. Mind-boggling, isn't it?

Welcome to the peculiar topic of **Quantum Superposition and Measurement**. This is where our classical expectations take a backseat, and the quantum realm reveals its surreal rules.

Superposition: The Heart of Quantum Mechanics

At the core of quantum mechanics is the principle of superposition. It states that a microscopic particle, such as an electron, can exist in multiple states or locations simultaneously. It's a fundamental concept, like bits in a computer that can be either **0** or **1**, quantum bits—or qubits—can be in superpositions of both **0** and **1** at the same time.

The iconic Schrödinger's Cat thought experiment beautifully illustrates this. Put simply, Schrödinger devised a scenario where a cat, a flask of poison, and a radioactive atom are enclosed in a box. If the atom decays, the poison is released, and the cat dies. If not, the cat lives. According to quantum mechanics, the cat is both dead *and* alive until an observer opens the box. Strange? Absolutely. But it sets the stage for understanding the quantum world where our everyday rules do not apply.

Quantum States: Embracing the Uncertainty

Quantum particles are described by wave functions—a set of probabilities that define the various states a particle could occupy. It's like an electron's personal dossier, containing information about all the potential outcomes you could expect if you decide to measure its position or momentum. But here's the trick: you never know which outcome you'll get until you perform that measurement.

Experimental evidence for superposition comes from phenomena such as electron diffraction. When a beam of electrons passes through two closely spaced slits and onto a screen, it creates an interference pattern, as if each electron went through both slits at once. It's compelling evidence that particles at the quantum level exist in multiple states at the same time.

Wave Function Collapse: The Observer's Conundrum

Now, what happens when we decide to peek at the quantum system—say, to check which slit an electron actually goes through? Well, the act itself alters the outcome, collapsing the wave function. This collapse pins down the electron to a single state, a single position. The interference pattern vanishes as if the electron's multiple potentialities were reduced to one definitive reality by the mere act of observation.

This phenomenon has provoked heated debates among physicists for decades. Is it the observer's consciousness that precipitates this collapse? Or is it the interaction with a measuring device that disrupts the superposition? The answers are still debated, and it's one of those tantalizing questions that keeps the curtain drawn on the theatre of quantum mechanics.

The Measurement Problem: A Thorn In the Side

Enter the measurement problem: a fundamental and inescapable issue in quantum mechanics regarding the transition from the quantum world to the reality we observe. When exactly does the wave function collapse, and under what conditions? And how can one state or outcome be selected from among the many possibilities contained in the superposition?

This is more than a mere intellectual curiosity; it has real implications for quantum technologies. Understanding measurement is critical for developing quantum computers, where qubits must be measured without disturbing their fragile superposition states too early in the computation process.

Quantum Zeno Effect: Halting Time with Observation

An intriguing outcome related to measurement is the Quantum Zeno Effect. It's like the RAF's mythical Dambusters of World War II, who aimed to halt the production of German power by destroying their hydroelectric dams—in our quantum analogy, frequent measurements halt the evolution of a particle's state. A quantum system that is observed continuously will not change. Time seems to stand still!

This effect has potential practical applications for quantum computing, where it could be used to preserve the state of qubits. But it also adds another stroke to the bewildering portrait of the quantum world—a canvas where time can be paused by the mere act of looking.

So, What Does This Mean for Us?

At this point, you might wonder: Does the weirdness of superposition and measurement have any bearing on our day-to-day life? The surprising answer is yes.

Every modern electronic device, from smartphones to MRI machines, owes its existence to the principles of quantum mechanics. The ability to control electron states in semiconductor materials has been pivotal for the development of computer chips, solar cells, and sensors. We're using quantum phenomena to advance our technology without even realizing it.

Looking into the future, quantum superposition will be the engine driving quantum computers, which will revolutionize industries with their capability to solve complex problems ranging from drug discovery to climate modeling.

Understanding this quantum world is akin to stepping through a looking-glass into a realm where intuition is challenged, and possibilities are endless.

It's a universe within our own—an invisible frontier that's everywhere around us, within every piece of matter, in every beam of light. By grasping the essence of quantum superposition and the curious case of measurement, we're unlocking the doors to understanding the universe and harnessing its laws for the next leap in human innovation.

And with that, you now hold the key to a brave new world that operates on the smallest of scales but promises immense possibilities. Embrace the quantum coin in your mind, with both its heads and tails in superposition, for it represents the doorway to a revolutionary future that's yours to discover.

Welcome to the wonderland of quantum improbability, where impossibilities collapse into reality, and where, with the knowledge you're gaining, you're not just an observer but also a participant in an extraordinary adventure. Keep learning, keep measuring, and most importantly, keep imagining.

BONUS CHAPTER: QUANTUM TUNNELLING

FIND OUT EVERYTHING YOU NEED TO KNOW ABOUT QUANTUM TUNNELING, INCLUDING HOW PARTICLES CAN PASS THROUGH BARRIERS.

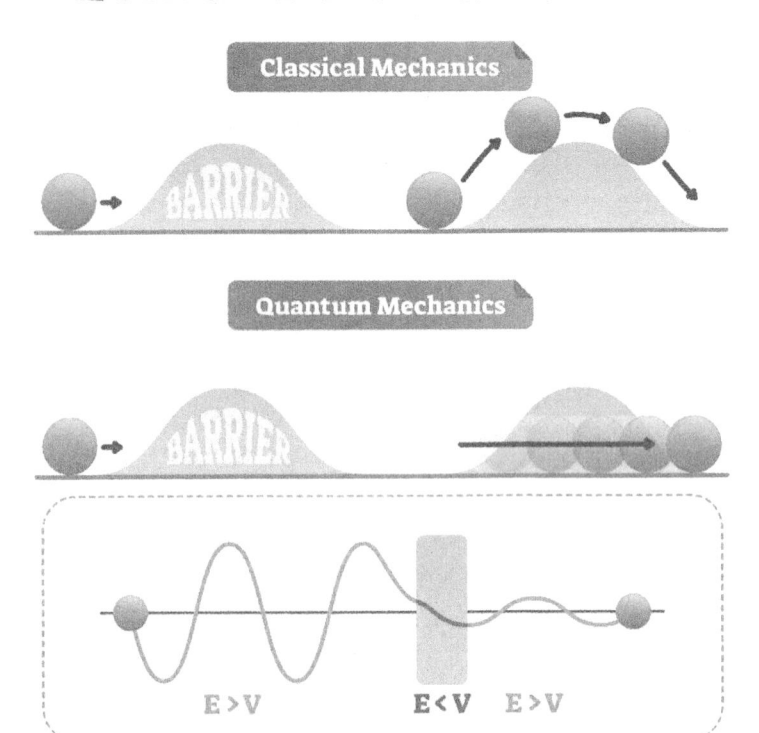

Imagine you are holding a basketball. Now, think about shooting it at a solid wall with the hope that it will pass through to the other side without demolishing the wall or bouncing back. In our everyday macroscopic world, this scenario seems like a page ripped out from a fantasy novel. Yet, in the bewildering quantum realm, particles achieve an analogous feat known as quantum tunneling. It's a phenomenon that not only defies our classical intuition but is also one of the cornerstones on which our modern technological edifice is built.

Quantum tunneling describes the process by which particles move through a barrier that, according to the physics of our tangible world, should be impenetrable. This doesn't mean that particles grow superpowers or that the barriers suddenly vanish; rather, particles take advantage of the probabilities that quantum mechanics grants them.

The concept originates from the wave-particle duality of particles. Much like how a wave can spread out and infiltrate spaces where a solid particle cannot, quantum entities like electrons exhibit wavelike characteristics. Thus, when confined by a barrier, an electron's wave function—the mathematical representation of its quantum state—does not abruptly stop at the wall's edge. Instead, it gradually diminishes, but a tiny part seeps into and beyond the barrier, appearing on the other side.

This occurrence is radically counterintuitive because it seems to violate the conservation of energy. Hypothetically, if an electron does not carry enough energy to surmount the barrier, how can it possibly appear on the other side? The answer lies in the probabilistic nature of quantum mechanics—a framework where the impossible becomes merely improbable.

Let's put aside the notion of particles as pinpoints navigating space, and instead envision them as a haze of potentialities—a cloud where the electron might be here, there, or even on the opposite side of a barricade. When a particle confronts a barrier, there's a certain probability that it will tunnel through and emerge on the other side.

The likelihood of tunneling is determined by multiple factors: the energy of the particle, the width and height of the barrier, and the mass of the particle. Thinner barriers and higher energies yield a greater probability of tunneling. Moreover, the mass of the particle holds sway over its capability to tunnel. Smaller particles, such as electrons, are much more proficient tunnelers than their bulkier counterparts.

Skeptical? You might be surprised to learn that quantum tunneling is not a mere theoretical construction. It has been experimentally observed and is, in fact, a phenomenon that is essential to several aspects of our daily lives. It fuels the nuclear reactions in stars, including our own sun, where the blistering temperatures and pressures still fall short of the requirements for nuclear fusion without the aid of tunneling. It is also seminal to the operation of our very own biological machinery; for instance, mutations in DNA sometimes occur as a result of protons quantum tunneling from one position to another, altering the genetic code.

In the technological sphere, tunnel diodes exploit this phenomenon for their ability to operate at incredibly high speeds, making them valuable in radio frequency applications. Scanning tunneling microscopes, which can image surfaces at the atomic scale, rely on the tunneling of electrons to generate pictures with astonishing precision. Furthermore, quantum tunneling is a critical factor in modern electronics, particularly in the operation of transistors. As these components continue to shrink with the relentless march of Moore's Law, tunneling effects become more pronounced and pose both challenges and opportunities for the industry.

But quantum tunneling is not without its pesky aspects. In some cases, it can be an unwelcome guest. For instance, it is among the issues in creating sustainable quantum computing systems. Quantum bits, or qubits, are susceptible to losing their quantum information through tunneling events - a nuisance known as quantum decoherence.

Despite these challenges, quantum tunneling also opens up vistas of potential applications. One possible futuristic application lies in the development of tunneling transistors, which could lead to computers that are not only exponentially faster but also more energy-efficient. In another ambitious application, researchers are exploring the use of quantum tunneling in solar cells to improve their efficiency well beyond what is possible with traditional photovoltaic technology.

Even more spectacular is the theoretical proposal of "quantum batteries" that use tunneling to enhance charging processes. These ideas, though presently confined to the research arena, tantalize with the promise of a leap forward in energy storage technology.

The effects of quantum tunneling may even extend to the brain, where it could play a role in the mysterious and contentious question of consciousness. Some theorists propose that quantum effects in neuronal structures contribute to the emergence of thought and awareness, an area of research that invites both excitement and skepticism.

All these facets suggest a landscape where quantum tunneling is not merely a curious quirk of the quantum world; it is a pivotal player in the phenomena that define our universe and a versatile tool in the evolution of technology.

Attempting to provide a comprehensive explanation of quantum tunneling is akin to trying to describe a vast tapestry in a few brushstrokes. Yet, what can be said with surety is that this phenomenon embodies the essence of the quantum domain—a place where reality defies common sense and where particles act in ways that would leave Newton scratching his head in utter bewilderment. Quantum tunneling reminds us that at the heart of the cosmos lies a strange and enigmatic quantum code—a principle that is not only intrinsic to the structure of the universe but is also increasingly integral to our manipulation and understanding of it.

James Philips writes with the intention of whisking you away from the grasp of esoteric equations into a narrative where quantum mechanics is not a stranger, but rather a fascinating companion. As you journey deeper into the scientific saga of quantum tunneling, remember that you are engaging with one of nature's most splendid riddles, a puzzle that has much more awaiting discovery and application.

In a world ever eager for smaller gadgets, more efficient energy solutions, and a deeper understanding of cosmic and human-scale processes, quantum tunneling stands out as a marvel that is both relevant and rich with potential—a thread in the fabric of existence that exemplifies the quantum dance of possibilities.

SCAN QR CODE TO DOWNLOAD FREE AUDIOBOOK VERSION

Made in the USA
Coppell, TX
24 August 2024